Tourism: A Temporal Analysis

Edited by Philip Goulding

(G) Goodfellow Publishers Ltd

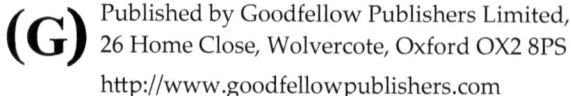 Published by Goodfellow Publishers Limited,
26 Home Close, Wolvercote, Oxford OX2 8PS
http://www.goodfellowpublishers.com

British Library Cataloguing in Publication Data: a catalogue record for this title is available from the British Library.

Library of Congress Catalog Card Number: on file.

ISBN: 978-1-911635-85-7

DOI: 10.23912/9781911635840-4702

Copyright © Philip Goulding, 2023

All rights reserved. The text of this publication, or any part thereof, may not be reproduced or transmitted in any form or by any means, electronic or mechanical, including photocopying, recording, storage in an information retrieval system, or otherwise, without prior permission of the publisher or under licence from the Copyright Licensing Agency Limited. Further details of such licences (for reprographic reproduction) may be obtained from the Copyright Licensing Agency Limited, of Saffron House, 6–10 Kirby Street, London EC1N 8TS.

All trademarks used herein are the property of their repective owners, The use of trademarks or brand names in this text does not imply any affiliation with or endorsement of this book by such owners.

 Design and typesetting by P.K. McBride, www.macbride.org.uk

Cover design by Cylinder

Printed by Lightning Source

Contents

Preface v
About the editor and contributors viii

Part 1: Temporal Dimensions and Relationships

1 **Time and Tourism** 3
   Philip Goulding
   Case study: Tourism seasonality in Yorkshire 12

2 **Causal Factors of Seasonality and Temporal Imbalances in Tourism** 16
   Philip Goulding
   Vignette : 112 Reasons to trade seasonally 24

3 **Tourism and the Seasons** 29
   Adele Doran and Peter Schofield
   Case study: Winter sun - spring break in Miami 35
   Case study: Climate change and ski tourism 38

4 **Nature and Time** 43
   Adele Doran and Seonyoung Kim
   Case study 1: Whale watching tourism, Australia 47
   Case study 2: Dark sky tourism, UK 48
   Case study 3: Polar bear viewing tourism, Canada 49

5 **A Chronological Exploration of Key Influences on the Development of Tourism** 56
   Neus Crous-Costa, Dolors Vidal-Casellas and Nuria Morere-Molinero
   Case study : The continuity of expeditions 60

6 **Night and Light : Nocturnal Tourism** 71
   Raquel Cambrubí, Lluis Coromina and Jaume Guía
   Case study 1: Llum Barcelona 76
   Case study 2: London Night Time Commission 79

Part 2: Operational Dimensions of Temporality

7 **Temporal Pricing in Tourism** 85
   Natalie Haynes and David Egan
   Vignette: The ethics of personalised pricing 92

8 **Measuring Temporal Performance in Tourism** 95
   Natalie Haynes and David Egan
   Vignette: CMA investigation into booking site practices 102

9	**Planning for Seasons: the Macro Level** *Jana Heimel*	107
	Case study: SbB – Stuttgart by Bike	111
10	**Seasonal Employment in Tourism** *Tom Baum, Tara Duncan and Deborah Forsyth*	119
	Case study 1: Tofino, British Columbia, Canada	123
	Case study 2: Sälen, Sweden	127
11	**Temporality and the Lifestyle Operator** *Claire Holland*	131

Part 3: Strategic Responses to Temporality

12	**Temporality: the Destination Management Perspective** *Jean Metcalfe and Paul Fallon*	151
	Vignette 1: Destination marketing during a pandemic	156
	Vignette 2. Tackling seasonality	160
13	**Temporal Event Tourism Strategies** *Mark Norman*	166
	Case study: Tramlines in Sheffield	175
14	**Marketing the Seasons** *Richard Tresidder and Emmie Deakin*	180
	Case study: The National Trust at Calke Abbey	181
15	**Planning for Seasons: Value Chain Management and Digitization** *Jana Heimel*	191
	Case study: SbB – Stuttgart by Bike	193
	Case study: AIDA cruise line operator	195

Part 4: Covid and Post-Covid: Temporality Futures

16	**The Growth of De-Temporalisation in Tourism** *Alisha Ali and Philip Murray*	205
	Vignette: Destination Wow	216
17	**Seasonality and Overtourism** *Richard Butler*	220
	Case study: Venice – conflict and crowding	228
18	**Endnote: Covid and Beyond Covid: Temporal Futures** *Philip Goulding*	234
	Index	241

Preface

The middle of the 20th century saw the first academic publications focusing on the study of tourism. They emerged to support the newly established tourism, travel and spatial geography related courses appearing in academic institutions in various parts of the world. Over the intervening decades, as the tourism phenomenon has grown exponentially and become a major economic and developmental force globally, where 'everywhere' became a destination, so too has our understanding of its scope, its spatial dimensions, the driving forces behind its growth in a globalised world, its cultural, environmental and sociological impacts and the planning and management approaches adopted to realise sustainable tourism.

A common thread through much of the knowledge base has been to understand the factors influencing the demand for tourism. One aspect of this is to understand why people travel *when* they do, for touristic purposes. In the 1960s, bodies such as the IUOTO (now the UNWTO), IATA, ICAO, the OECD and a number of research institutes were already capturing tourism's demand dynamics through 'seasonality' measures, hence temporalizing tourism movements. In 1975, Raymond Bar On published the findings of an extensive longitudinal study of temporal performance variations in tourism across a number of countries, highlighting and analysing numerous dimensions of the phenomenon he referred to as 'seasonality'. A few years later, in the seminal anthropological study of tourism, Valene Smith introduced us to the impacts on host societies from seasonal influxes of visitors in various parts of the world: an economic 'feast or famine' situation in some places and a relief to some societies during the seasonal 'down-times'. The evolving literature informed us that temporality is both a demand-fuelled and supply-side determined phenomenon. Tourism is constructed around temporal factors both within and beyond the control of the sector and its participants, and thus has implications for consumers, businesses, destinations and the wider environment in which it operates. This book addresses a number of those interrelationships.

The book owes its genesis to the first tourism text solely devoted to the study of temporality: Tom Baum and Svend Lundtorp's edited tome *Seasonality in Tourism*, published in 2001. That was the first to address a wide range of both supply and demand factors to assess, through empirical studies, the characteristics, causation, performance and policy implications of seasonality, largely in cold water Atlantic destinations in which seasonal variations were marked.

'Seasonality mitigation' has long been a policy priority in many places and at numerous levels (local to inter-regional) in the face of increasing over-tourism and anti-tourism sentiment, while the concurrent emergence of digital technologies has enabled operators to use 'time' as a key factor in demand and capacity management, distribution and pricing. Forms of tourism predicated on time have long existed – winter sports tourism being the most obvious – and have subsequently emerged and developed into significant activities: the 'calendarisation' of tourism around, for example, cultural festivals, nature and natural phenomena; also nocturnalism, slowness and wellness among others. The time is right, arguably overdue, therefore, to revisit the inter-relationships of tourism and temporality, beyond the lens of 'seasonality'.

The purpose of this book is essentially fourfold. First, it brings together many of the threads and dimensions of temporality in tourism into a single volume and grouped into a number of themes, as discussed below. Importantly also, it extends the domain of temporality beyond 'seasonality', by adding other 'time dimensions' (chronology, periodicity, nocturnal/diurnal etc..) into the mix. In so doing, the book aims to redress the temporal element of the 'spatio-temporal' relationships in tourism, in which in the idea of 'mobilities' has subsumed 'temporalities'. Finally, the chapters embrace both conceptual and practical approaches to their respective topics, including short case studies or vignettes to illustrate their themes.

It is acknowledged that some dimensions of temporality could support a full chapter in their own right. For example, while 'fast vs slow' tourism is discussed within Chapter 4, 'slow tourism' has developed as a distinct body of knowledge within the wider international literature. Similarly, community responses to seasonal/periodic variations in tourism are acknowledged, for example in coping with over-tourism in Venice (see Chapter 16), though there is scope to bring together the many threads of that inter-relationship.

Since the genesis of this book, the Coronavirus pandemic of 2020-2021 has created a temporally more complex environment for tourism at all levels. 'Virtual tourism' was boosted by 'stay-at-home' experiences, some of which were time-specific, others less so because of streaming technologies and the capabilities of social media. Many of the chapters of this book were written during the early stages of the pandemic, in which 'travel shut-down' prevailed. They therefore reflect a pre-pandemic reality, while some chapters have made reference to a changing reality, in particular those focusing on operational aspects of tourism and travel business.

As mentioned above, the book is constructed in four thematic sections, each containing a number of chapters around that theme. Part 1 introduces the reader to a number of temporal dimensions and establishes a conceptual base for understanding the inter-relationships. Part 2 focuses on operational dimensions of temporality in tourism while Part 3 takes a more strategic focus. The final chapters, in Part 4, explore whether temporality is still a relevant concept, given the development of both tourism and digital technologies.

The book is designed so that each chapter can be read as a self-contained reader on its topic. However, some chapters can be read as 'themed pairings' (those by Doran, by Heimel and by Haynes and Egan) in which the topics of each naturally complement the other by the same author(s). Each chapter also contains a caselet / vignette to illustrate the conceptual content. In some cases these are also self-contained within the chapter; elsewhere (the chapters by Holland and by Heimel) the case 'runs through' the chapter narrative.

The editor is indebted to each chapter author and co-writer for their involvement in the book and their contribution to extending the knowledge domain of temporality and tourism. Their patience and forbearance with the extended time frame of this project, from genesis to end point is much appreciated. Thanks go also to all the individuals who have previously co-written with this book editor and/or have in the past provided academic direction in one form or another: Professors Tom Baum, Richard Butler and Brian Hay in particular. Finally, the unlimited patience, guidance and support from Sally North of Goodfellow ensured this book came to fruition. Thank you Sally!

Philip Goulding
  Until recently, Sheffield Hallam University

About the editor and contributors

Editor

Philip Goulding was Principal Lecturer in Tourism Management at Sheffield Hallam University until January 2023, when he retired after nearly 40 years in tourism academia. He previously worked at Edinburgh Napier University and at the former South Glamorgan Institute of Higher Education, after starting his working life in the travel industry in the UK and Iran. Along the way he was instrumental in the development of tourism management degree programmes and undertook various consultancy projects for national and regional tourism and destination management bodies in Scotland and Wales, often around the topic of seasonality mitigation. His doctorate at the former Scottish Hotel School (University of Strathclyde) was a study of the temporal trading behaviours of small tourism businesses in Scotland. Over the years he has published a number of journal articles and book chapters on various aspects of temporality in tourism.

Contributors

Alisha Ali is an Associate Professor and Head of Research Degrees at the Social and Economic Research Institute at Sheffield Hallam University. She is an interdisciplinary and internationally recognised researcher studying the sustainable development of tourism and hospitality information and communication technologies, social responsibility, entrepreneurship, working conditions and hospitality and tourism education. She has a sustained record of research activity published in peer-reviewed journals, books, book chapters, conference papers, book reviews and trade press. Alisha also has a background in consultancy, working with government offices, destination management organisations (DMOs) and international, national, and local businesses. She has delivered invited talks worldwide and has extensive experience in PhD supervision, completions and examinations.

Thomas Baum is Professor of Tourism Employment at the University of Strathclyde, Glasgow, and previously Head of Department of HRM and the Scottish Hotel School. He has spent the last 40 years dedicated to the study of the tourism and hospitality workforce as researcher, teacher and consultant. He has worked in 45 countries and provided policy-informing advice to governments, international agencies (including UN agencies) and the private sector on a range of issues in this area. He completed a second doctorate (DLitt) to bring together 35 of his key publications over 30 years in this field. He was co-editor of the first book dedicated to tourism temporality, *Seasonality in Tourism* (2001).

Richard Butler is Emeritus Professor of Tourism at the University of Strathclyde, Glasgow, and Visiting Professor at NHTV University, Breda, Netherlands. He is a geographer, with degrees from Nottingham (BA) and Glasgow (PhD) Universities. He taught at the University of Western Ontario and the University of Surrey, where he was Deputy Head (Research) at the School of Management. Richard has published 23 books on tourism and has authored over a hundred journal articles and chapters in books. He was awarded the UNWTO Ulysses Award in 2016 for "excellence in creation and dissemination of knowledge". He has served as consultant for government agencies in the UK, Canada and Australia and is on the editorial board of several tourism journals. His main areas of research are tourism destination development, tourism in remote areas, seasonality, and the sustainability of tourism. He is a founding member and former President of the International Academy for the Study of Tourism.

Raquel Cambrubí is Associate Professor at the Department of Business Administration in the Faculty of Tourism, University of Girona. She completed her PhD in 2009, focusing on tourism image formation and relational networks. Her research interests cover destination management, tourism image and branding, tourist behaviour, and risk perception. She published in several international journals such as *Annals of Tourism Research* and *Tourism Management*.

Lluis Coromina is Vice Dean at the Faculty of Tourism, University of Girona. He completed his PhD in 2006. His research is related with cross-cultural comparison, quantitative research methodology and tourism behaviour. He also publishes in related fields such as sociology or political science. He is reviewer of international journals, and his publications appear in *Tourism Management,* and *Current Issues in Tourism,* among others.

Neus Crous-Costa holds an MA degree in Tourism Planning and Management. She is professor and researcher in the field of tourism and personal development in the Department of History and Art History at the Universitat de Girona, Spain. Her other research and teaching interests include tourism as a tool for dialogue, heritage management and interpretation, spiritual tourism and cooperation.

Emmie Deakin is Senior Lecturer in Hospitality and Tourism Management at Sheffield Hallam University. Gaining her doctorate in Architectural Conservation, she has researched and published on various aspects of historic buildings and heritage management, hospitality experience-scapes and industrial heritage. She has been course leader for postgraduate programmes for Tourism and Hospitality and Student Experience Lead in the Department of Maths and Engineering at Sheffield Hallam.

Adele Doran is a Principal Lecturer in Tourism Management at Sheffield Hallam University. Her research focuses on outdoor recreation and adventure tourism, including the gendered experiences of participants, participant well-being, adventure media and user-generated content, charity challenge tourism and women's entrepreneurship and employment within the sector. Adele is a committee member of the Adventure Tourism Research Association (ATRA) and Sheffield Hallam University's Outdoor Recreation Research Group.

Tara Duncan is Professor of Tourism Studies in the School of School of Culture and Society at Dalarna University, Sweden. Tara has published extensively on topics addressing various aspects of temporality including seasonality and migrant workers in tourism, mobilities of work, second homes, e-tourism technologies and consumer behaviour, among others. She is a member of CeTLeR: Centre for Tourism and Leisure Research and is Chair and Co-ordinator of ATLAS (Association for Tourism and Leisure Education and Research).

David Egan is a Senior Lecturer in the Sheffield Business School at Sheffield Hallam University. He teaches in Hospitality Management and is an active researcher widely published in the areas of revenue management, the culture of cafes and various aspects of the economics of the hospitality and tourism sectors. David has over 30 years' experience of delivering consultancy type projects for a wide range of clients including the tourism sector focusing on economic impact studies and tourism destination management. He is also an experienced doctoral supervisor.

Paul Fallon is at the Lancashire School of Business and Enterprise, University of Central Lancashire, where he is primarily responsible for teaching at UG and PG level as well as being an International Lead. He has accumulated over 20 years teaching experience gained from teaching at four HEIs and their international partners (including in China, Singapore, Malaysia and Switzerland). His teaching foci include tourism, hospitality and event marketing. Together with Dr John Heeley, Paul developed the annual Destination Marketing Symposium at Sheffield Hallam University to enable students to learn directly from practitioners in different destination contexts. Paul's research interests are mainly centred around destination marketing, consumer behaviour and customer experiences. He is also a member of the ATHE Senior Executive team.

Deborah Forsyth is currently a doctoral candidate in the Department of Work, Employment and Organisation at the University of Strathclyde where she is engaged in research on the intersection of labour migration and employment. Her main areas of interest encompass broadly temporary labour migration and temporary labour migration programmes, low wage work, labour mobility, labour process and employment relations, rural economies, and the hospitality sector. Her current research focuses on how migration regulatory controls associated with employer sponsored temporary migration programmes for low-skilled migrant workers impact labour mobility and the effort bargain.

Jaume Guía is an Associate Professor at the Department of Business Administration in the Faculty of Tourism of the University of Girona. He is Programme Director of the Erasmus Mundus European Masters in Tourism Management (EMTM) and Scientific Director of the Tourism Research Campus at the university. His most recent publications in academic journals deal with various aspects of destination management, with emphasis on destination branding, destination governance and cross-border destinations. He is also currently doing research on place making and tourism, on new mobilities and forms of tourism, and in justice tourism.

Natalie Haynes is a Principal Lecturer in the College of Business, Technology and Engineering at Sheffield Hallam University where she specialises in teaching hotel and airline revenue management for the Department of Service Sector Management and supervises several doctoral students. Her research and publications focus on the topic areas of revenue management, pricing and big data with a special focus on the hotel sector. She completed her Ph.D. in 2019 using a Straussian grounded theory approach to explore the use of big data in hotel price decision-making.

Jana Heimel is Professor at Heilbronn University and lecturer at Hochschule der MedienStuttgart, teaching international business and management accounting, strategic and change management. She also has extensive experience as a management consultant. With over 30 publications to her credit, her research areas are very varied, currently marked by a clear focus on sustainable mobility and bicycle tourism. A keen cyclist and alpinist, she founded Stuttgart by Bike, a cycle tourism and event organization.

Claire Holland is a Principal Lecturer in the Department of Management at Sheffield Hallam University. Her research interests focus on mobilities and the role of work in a liquid world, the social context of work and the future of work with particular reference to the hospitality and tourism industries. Her teaching focuses on leadership, organisational behaviour and human resources management. Claire has a passion for developing student self-awareness to support learning.

Seonyoung Kim is a Principal Lecturer in Tourism Management at Sheffield Hallam University and previously Subject Group Leader of the Tourism team. She has been instrumental in the development of new courses in both tourism and aviation management. She researches tourism governance, tourism policy and planning, urban tourism, destination marketing, and accessible tourism.

Jean Metcalfe has extensive professional experience. Starting her career working for an independent tour operator and then a recruitment consultancy, she was appointed as Public Relations and Training Officer for Cumbria Tourism, working closely with public and private sector organisations to promote and manage the destination. Then as Director at Greater Glasgow Tourist Board and Convention Bureau, Jean had responsibility for all Visitor Services and for the successful completion of a £2 million Visitor Information Centre. Jean subsequently joined UCLan as a Senior Lecturer and Programme Leader in 1996 and has used this experience extensively in her teaching of tourism and personal development modules, and is now bringing it to bear on research.

Nuria Morere-Molinero is Director of the Masters programme in International Direction of Tourism at Rey Juan Carlos University, Madrid. Her research field focuses on the links between cultural heritage, history and tourism. This includes areas such as cultural routes, archaeological sites and tourist attractions, as well as the history of travel and tourism.

Philip Murray is a Senior Teaching Fellow in Operations Management in the Department of Business Transformation at Surrey Business School, where he teaches and supervises students on UG and PG programmes. Philip is a Fellow of the HEA and an experienced programme and module manager who has an active role in curriculum and new program development. Prior to joining academia, Philip worked in operations for international hotel brands like Crowne Plaza & Clarion Hotels, as well as SMEs in the tourism sector.

Mark Norman has been teaching event management and tourism at Sheffield Hallam University for the last decade. Prior to this, he had an extensive career in music and festival events. Today, he still works as a freelancer in the industry when time allows. Some of the past events he has worked on include the Olympic Torch Relay, Tour de France and Leeds Festival. He is also undertaking a PhD at the University of Tilburg in the Netherlands, investigating event tourism strategies in towns and cities.

Peter Schofield is Professor of Tourism and Services Management at Sheffield Hallam University and a founder member of the International Academy of Culture, Tourism and Hospitality Research. His research and consultancy interests include the consumer psychology of leisure and tourism, destination and events marketing, and services management. Within these areas, his core research is in consumer decision making and behaviour, compulsive purchase behaviour, tourism market segmentation, and service failure and recovery.

Richard Tresidder is Associate Professor in Marketing and Hospitality Studies in the Department of Management at Sheffield Hallam University and was Programme Leader for the Sheffield Business School DBA. Prior to joining

Sheffield Hallam, he held various positions at Roehampton, Anglia Ruskin, Derby and Keele Universities. He has published widely within the fields of hospitality and tourism, with a particular focus and research activity in the areas of semiotics, marketing, heritage management, tourism development, food in contemporary society and deviancy in the workplace.

Dolors Vidal Casellas works on cultural tourism from a political and managerial perspective, as well as the connection of culture with artificial intelligence and other topics such as spiritual tourism. Her PhD thesis demonstrates the existence of cultural tourism in Barcelona before World War I. Dolors is Director of the Gastronomy, Culture and Tourism Chair at Calonge – Sant Antoni.

Part 1:
Temporal Dimensions and Relationships

Any form of human activity is predicated on time. In the case of tourism, temporality is embedded in many different ways. The very act of travelling consumes time. The time periods and durations in which human movement for touristic purposes takes place between any two points create patterns which in themselves may give rise to issues or challenges for businesses and destinations involved. Individuals' time-related travel decisions may depend on a set of variables within the gift of the traveller or visitor. On the other hand, those decisions may be constrained by factors beyond their control or facilitated by temporal markers. Moreover, time-related phenomena may provide the basis for touristic activity. Therefore, the natural starting point for this book is to establish a conceptual base around temporality in tourism and explore some of the main intersections between the two phenomena, hence 'dimensions and relationships'. As such, the chapters within this first section provide a context for the various chapters in the subsequent three sections.

Chapter 1 uses a systems approach to help understand where and how temporality is embedded within 'the tourism system'. Our understanding of temporality in tourism has developed from the notion of 'seasonality', an endemic component of tourism. Therefore, the chapter also sets out to explore the concept of seasonality within a wider temporal framework. It demonstrates that 'seasonality' is not the only temporal construct and seeks to differentiate it from 'periodicity' or periodic temporal variations, in order to provide a foundation for operational and management issues explored in chapters later in the book.

Chapter 2 then focuses on unravelling causal factors that influence wider temporal and seasonal patterns and variations in tourism, building on existing conceptual frameworks. While much of the focus in the literature has traditionally been on demand-side causal factors or influences, this chapter takes a wider perspective, exploring supply-side and 'modifying' influences on temporal causation and questioning the relational boundaries between them.

The next two chapters each examine fundamental conditions related to temporality: the influence of the seasons and climatic variation (Chapter 3) and that of the natural world, i.e. nature and time (Chapter 4). Climate and weather events are a fundamental determinant of the temporality of tourism, aspects of which Chapter 3 explores. Nature-based tourism takes many forms, which in turn are predicated on aspects of time such as natural migrations, dark skies and relative speed. These are the focus of Chapter 4.

Another key construct of time is chronology. In Chapter 5, a chronological construct is used to help our understanding of how the development of tourism over the past few centuries has been informed by a range of historical events, including social, cultural, political and industrial advances. Finally, Chapter 6 focuses on the nocturnal/diurnal dimension and how in particular the phenomena of night and darkness have influenced the development of forms of touristic activities.

1 Time and Tourism

Philip Goulding

Learning outcomes

By reading this chapter, students should be able to:

1. Describe the phenomenon of temporality in tourism using a 'systems approach' to understand its implications.
2. Identify and differentiate terminologies associated with tourism 'seasonality'.
3. Identify and differentiate different types of tourism seasonality across the world.
4. Appreciate the construct of 'periodicity' in tourism related operations, as temporally distinct from 'seasonality'.

Introduction

The relationships between tourism and time are multi-dimensional and complex. At a macro level, the mass movement of people for touristic purposes within and across the various parts of the globe appears as a relentless surge of travel throughout the course of the year. However, in practice, the temporal spread of tourism is far from uniform. The reality is a complex mix of travel patterns that pulse in intensity at different times of the year according to geography, destination resources, climate, human motivations, personal circumstances, economic wellbeing, transport infrastructures and the activities of a sophisticated tourism industry that responds in numerous ways to temporal variations in demand. The shorthand for this is 'seasonality', a concept that has been synonymous with tourism ever since recreational travel became established (Butler, 2001; Bar On, 1975).

But while seasonality has traditionally been viewed as a dominant characteristic of tourism, it is by no means the only temporal construct that characterises the industry. Temporal variations in the demand for and supply of touristic services exist not only across 'seasons', but across shorter time spans at every duration: months, weeks, weekdays, weekends, public holidays, times of the day, diurnal and nocturnal periods, durations of festivals and of natural phenomena. This connotes 'periodic' variation (Hartmann, 1986; Goulding & Pomfret, 2022). Passenger transport and tourism businesses routinely construct their operational and strategic planning to accommodate and predict temporal variations in demand and supply over both longer (annual and several years in advance) and shorter (i.e. periodic) time frames.

The purpose of this chapter is therefore to establish a framework of analysis to help appreciate time-related dimensions of tourism. It acts as a platform upon which the following chapters will develop particular temporal inter-relationships. This chapter will explore the concept of 'seasonality' and why it has been viewed as a challenge or 'problem' to be 'overcome' at destination level. Following on from this, the concept of 'periodicity' in tourism is examined, illustrating its implications in a number of operational contexts. Finally, a short case example of tourism seasonality in a UK region is examined.

An overview of temporality in tourism

In order to aid our understanding of how temporality, expressed as the various dimensions of time, is inherent within tourism, we can take a 'systems approach' based on the idea that tourism itself is a system of inter-related constructs and components (see for example Leiper, 1990, Page et al., 2001). In other words, temporality is a subset or dimension of the tourism system and which has consequences throughout that system.

First, consumer demand is temporally driven to a significant extent. For example, consumer market segmentation acknowledges 'time poor' consumers who may pay premium rates to travel within restricted time periods or where 'time is of the essence' in reaching a destination as quickly as possible. Conversely some consumer markets are 'time rich' and can adjust their travel to achieve best personal outcomes (eg lowest cost, greatest availability of services to choose between). Transport networks adjust pricing to reflect such temporality in demand. (This is explored in Chapters 7 and 8 of this book). Likewise destination services including accommodation are very often sensitive to temporal demand variations. Hence, there is a series of relationships in matching temporal visitor demand and business operations, as demonstrated by the Stuttgart by Bike case example in Chapters 9 and 15.

Temporal imbalances can impact significantly on destinations' carrying capacity, their human, natural and environmental resources. This may be the consequence of significant temporal demand variations (such as in the case of peripheral 'cold water' destinations in the North Atlantic margins (Baum & Hagen, 1999)) or where demand broadens out across a year, resulting in extreme though increasingly common cases of over-tourism even within relatively less-visited times of the year (as illustrated in the case of Venice in Chapter 17).

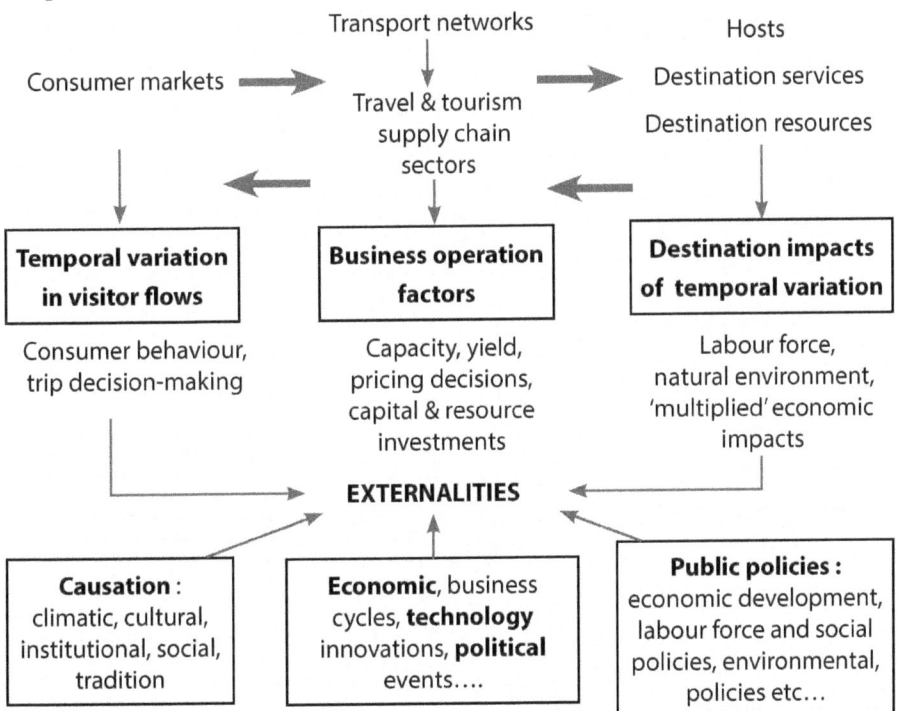

Figure 1.1: A systems approach to tourism temporality. Source: Author

The responses that destinations (their communities, pressure groups, local, regional or national governments or public authorities) are able to make to the challenges of temporal fluctuations are therefore aimed towards mitigating externalities. Responses may be through public policies, marketing or addressing specific causal factors, in order to achieve temporal sustainability from tourism. Figure 1.1 summarises this 'systems approach' applied to temporality in tourism.

The construct of 'seasonality' in tourism

Whether at international or domestic levels, modern-day tourism has tended to evolve from a complex inter-play of temporal factors that has shaped the mass movement of people. Travel for trade, for pilgrimages and later for cultural enlightenment (as discussed in Chapter 5) has for centuries occurred in temporally defined periods. Since the nineteenth century, mass leisure travel in Europe and North America has significantly evolved on the basis of recreational activities that are seasonally defined, leading to the development of resorts in 'summer sun' and 'winter sun' coastal destinations (typically in the Caribbean, Florida, Mediterranean) and winter sports-based tourism destinations (in the Rocky Mountains, European Alps and Pyrenees, for example). Within the last half-century, distinct 'summer' and 'winter' inclusive tour operations have defined spatial flows of leisure tourism from the cold and cool temperate climes to the warmer, Mediterranean climes. Thus was penned the concept of 'seasonality' to characterise such movements (Haulot, 1963; Boyer, 1972; Bar On, 1975).

In essence, seasonality is a pattern of movements during a particular time-period within the course of a year that recurs on an annual basis (Frechtling, 2001). This definition is significant given the need by economists and tourism authorities to forecast the temporal patterns of the volume, value, scale and impacts of tourism movements. Historically, such movements of people were seldom balanced across all months of the year. Tourism 'seasons' have thus connoted periods of time that characterise 'peaks' or particularly busy periods in travel and destination services. Tourism seasonality is therefore traditionally framed as representing temporal imbalances in the demand for tourism services. These imbalances are recorded through a range of measures in which individual months tend to be taken as the standard seasonal unit of measurement (Duro & Turrión-Prats, 2019). However, tourism seasonality data is often aggregated to three-monthly (i.e. quarterly) periods, in order to simplify the task of year-on-year trend analysis.

Butler's (2001: 5) definition of seasonality identifies that temporal imbalances in tourism:

'may be expressed in terms of dimensions of such elements as numbers of visitors, expenditure of visitors, traffic on highways and other forms of transportation, employment and admissions to attractions'.

In other words, seasonality involves both volume and value indicators of tourism activity. Previously, Butler reviewed a longitudinal study of volume data of temporal variation measures on tourism in Scotland (Butler, 1994). He analysed 67 data sets including air, rail, sea and road transport passenger numbers, bridge and ferry crossings, employment data, accommodation occupancy and visitor attractions (specifically entrance figures to selected National Trust for Scotland (NTS) properties. From such data, he was able to construct indices of seasonality, reflecting the ratio between peak (highest) and trough (lowest) month tourism activity for the years 1970, 1980 and 1990 for areas within the Scottish Highlands. His study had significant empirical value, being case specific (the North and West Highlands of Scotland being areas exhibiting significant seasonal concentrations in their tourism economies) and for the insights into methodological issues and problems in quantifying seasonal disparity.

According to Grant, Human and Le Pelley (1997), the phenomenon is observable as peaks and troughs of visitor numbers during a calendar year. As will be demonstrated in ensuing chapters in this book, such peaks and troughs, if significant, can give rise to a series of problems for destinations and service providers alike and more broadly for the natural environment.

It should be noted that aggregated destination measures of seasonality (e.g. at country, regional or city level) may hide disparities among individual service providers. If seasonality represents market performance, this clearly is likely to vary significantly according to the composition and characteristics of a business' market. For example, city centre hotels geared to business travellers may have distinctly different patterns of demand compared to those of visitor attractions within the same locality, whose customer base may include local residents and educational parties. There is also the dimension of whether seasonal disparities apply equally to outbound vs inbound tourism activity. Peaks of outbound international travel movements from ports and airports in origin markets at key times of the year (high summer, school holidays, in the days leading to major holidays or religious festivals, for example) are associated with similar peak patterns in receiving countries or regions. However, the patterns may not necessarily match, given the disbursal of outbound travel to multiple destinations as well as the temporal characteristics of the various origin markets and travel motivations of visitors to a particular destination. For example, Thailand's peak visitor arrival months are typically November till January, while UK visits to the country are dispersed more widely across the year, depending on market characteristics, with younger travellers more summer oriented (UNWTO 2021).

Apart from peaks and troughs in visitor movements and earnings, other characteristics of temporal imbalances have led to variations in terminologies. For example, the 'shoulder' period is a term frequently used to describe the period between peak demand and least demand (Beaver, 2005). It is vague in the sense that the characteristics of the shoulder period are not well defined and can in themselves vary in duration. This vagueness equally applies to other temporal terms often used within the travel industry, such as 'high season', 'mid-season', 'low season' and 'off peak'. Other seasonal nomenclature identified in the literature (Butler & Mao, 1997; Chen & Pearce, 2012) includes 'twin peaks', 'multi-peaking', 'plateauing' of demand over a time duration or, as is often the case in large urban areas with mixed markets, a long history of hosting tourism and a mature tourism infrastructure, 'non-peak' tourism. This latter may make 'shoulder' or 'low season' an obsolete concept in those places (eg London, New York).

Chen and Pearce (2012) identified and classified a number of distinct tourism seasonality patterns in countries, based on their tourism arrival patterns, as follows in Table 1.1. A subsequent study by Duro and Turrión-Prats (2019) identified countries exhibiting such seasonal profiles.

Table 1.1 : Types of tourism seasonality

Type	Description
Rolling Hills	A series of continuous peaks and troughs throughout the year, above and below the monthly average. Examples: Brazil, Mexico, Thailand.
Single Peak Mountain	In which a steep single increase in arrivals peaks for a short period to a level significantly above the aggregated monthly average, before descending to a low trough. Examples: Bulgaria, Croatia, Finland.
Multi-peak Mountains	In which two or three distinct peaks in arrivals are offset by corresponding troughs. Examples: Austria, Dominican Republic.
Plateau	Where above average demand for several months of the year fluctuates with its own mini peaks and troughs, before falling to a pronounced low for the remainder of the year. Examples: Czechia, Japan, Netherlands.

Since tourism strategies and destination-wide collaborative initiatives often involve interventions and campaigns promoting 'low-season' market growth, the need for terminological clarity in understanding the temporal profile of a destination's tourism is evident. More recently, and as is demonstrated later in this book, temporal profiling is important for knowing when during the year it is necessary to suppress the volume of tourism.

'Periodicity' in tourism

While we have so far considered tourism's temporal imbalances from a seasonal perspective (i.e. variations of visitor movements from month to month or longer timeframes), this is only part of the full temporal equation. For most tourism businesses, demand for services fluctuates significantly within short time frames, typically within the course of a day, from day to day and between weekdays and weekends. Such short-term temporal variations can be described as 'periodicity' (in contrast to 'seasonality'). Periodic demand fluctuations have a more immediate impact on the day-to-day operations of a wide range of businesses including hospitality operations, transport, visitor attractions and leisure facilities (Frechtling, 2001; Su & Wall, 2016).

According to Frechtling (2001), monthly fluctuations in tourism volumes and earnings are largely predictable from year to year. There is a caveat to this, however. There are many events and festivals that do not occur annually, either because they are staged in longer term cycles and in different locations every time (Fifa World Cup, summer and winter Olympics and the UN 'COP' climate change conferences being among the more high profile, globally) or one-off / *ad hoc* events within destinations, that generate significant influxes of visitors within relatively short time periods. Examples of this latter include special temporary 'flagship' exhibitions in museums and galleries (the Terracotta Warriors world tour), world-tour concerts by leading rock bands and the launch of a new facility such as a conference centre or heritage attraction. The visitor demand they generate will typically distort the temporal balance of tourism in the locations where they are held for the duration of the event.

Moreover, the number of weekends in a month and the phenomena of annually movable but time-specific festivals will add to the distortion of short-term periodic performance, as measured by visitor movements and arrivals. Examples include the two week *Now Ruz* (Iranian New Year) period, the 10 day 'high holiday' period over *Rosh Hashanah* and *Yom Kippur* (Jewish New Year period), and Easter. These are **calendar effects** (Bar On, 1975). Other short-term irregularities such as a one-off conference will also have a periodic effect on local visitor numbers, hotel occupancy and revenue.

There is a debate as to whether 'periodic' fluctuations in tourism activity should be treated separately from 'seasonality', whether for measurement purposes or for appraising their effects (economic, social and environmental) (Getz and Nilsson, 2004; Su and Wall, 2016). Accordingly, the temporal measurement of tourism by government agencies, local tourism marketing or economic development agencies typically focuses on monthly or quarterly data rather than fluctuations between midweeks and weekends, week by week variations. Table 1.2 provides an illustrative summary of dimensions of short-term periodic temporal variation.

Table 1.2: Examples of periodic temporal variations in tourism related businesses

Periodic perspectives	Examples
Morning, daytime or evening-oriented service	Restaurants, takeaways, cafes, bars, bistros, coffee shops etc… where there are likely to be temporal pulses in demand across the course of a day or services restricted to certain hours to capture specific markets. Likewise concerts, shows (evening vs matinee)
Weekend-oriented service	Bars, nightclubs, some restaurants, many visitor attractions …. where service hours are more geared towards weekend recreational activity
Nocturnal events	Son et lumière events, winter-night festivals, night-sky phenomena and events (explored in Chapter 6)
Diurnal peak vs off-peak services	Transport operators such as airlines, trains, ferries, who set their schedules to reflect passenger demand and yield
Fixed-period festivals and events	Music festivals, arts festivals, agricultural fairs etc… drawing in visitors over a period typically of a few days to two weeks (see Chapter 14)
Late opening/early closing	Visitor attractions, retail centres, museums etc., reflecting operational necessities (staff training, stocktaking etc.)

Periodic variations in operating may reflect distinct visitation patterns by different market segments, for example between local populations and non-local visitors. Su and Wall (2016) undertook a study of visitor patterns by local residents, non-local domestic visitors and international visitors to a World Heritage site in Beijing. It revealed that local residents' spatial use patterns, dwell times, interactions and their visitor experience differed from those of visitors from further afield. Quite simply, they valued 'avoiding the crowds' by arriving earlier in the morning and later in the day. Therefore, by knowing the periodic fluctuations in demand for their services, tourism and hospitality operators can adjust their operating times and can act upon any visitor experience issues for the different segments.

From a destination-wide perspective, very short-term periodic patterns of tourism activity (eg diurnal/nocturnal, weekday vs weekend) are considered to be generally less stable than seasonal patterns (Lundtorp, 2001; Frechtling, 2001) although they may form part of a more definable longer-term cyclical pattern of temporal variation.

Operationally, periodicity is associated with fluctuations in daily work patterns and utilisation of facilities. During peak seasons in the year, this is amplified (Getz et al., 2004). For small tourism enterprises in particular, intense activity over a longer period may impact on their trading behaviours and operating decisions in quieter (off-peak) times, especially the desire for periodic rest and relaxation (Goulding, 2006; Connell & Page 2015).

In summary, an awareness of periodic patterns of demand aids an understanding of temporal cycles in tourism that may be seen as distinct from longer term patterns of seasonality.

Case study: Tourism seasonality in Yorkshire

As one of England's largest counties, Yorkshire is home to historic urban centres, national parks, seaside resorts, maritime and heritage attractions, renowned events and centres of commerce, sports and entertainment, hence a very diverse tourism product. With such a broad market composition, it could be assumed that seasonal demand disparity would not feature as a characteristic of the county's tourism economy. However, a range of data collected in 2019 and 2020 by Welcome To Yorkshire (WTY), at that time the county's tourism marketing and development agency, demonstrates the issue. In Figure 1.2, the temporal spread of visits and visitor spend are two fundamental measures used by WTY to illustrate how 'seasonality' is manifested in the tourism economy. The distributions represent quarterly percentages of the annual totals.

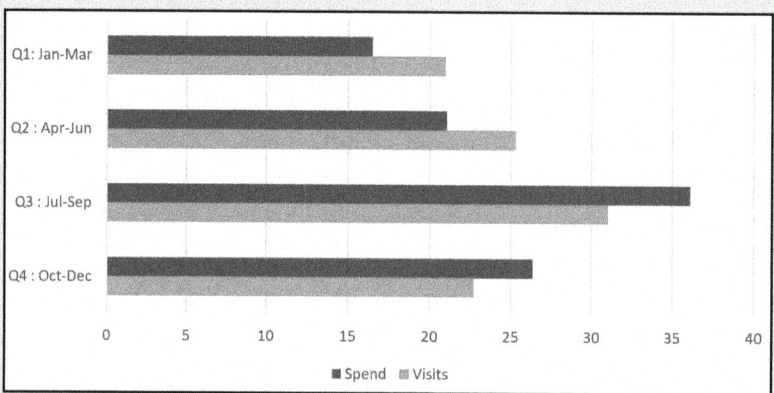

Figure 1.2: Quarterly distribution in volume of visits and visitor spend in Yorkshire, 2018. Source: Welcome To Yorkshire (2020)

The annual 'headline profile' of tourism seasonality in Yorkshire is apparent when the data are divided into four quarters. The summer period (Q3: July-September) is the main period of visitation to Yorkshire followed by spring, a picture replicated in other regions of the UK and in many European countries (Eurostat, 2021). However, visitor spend can be seen to display more extreme seasonal disparity between the four quarters, comprising around 16.5% of the annual spend total in Q1 (broadly winter and early spring) but over 26% of the annual total in Q4 (autumn through to the year's end). In Q3 and Q4, the proportion of spend exceeded visits, in contrast to the first half of the year.

Using such aggregated data, market analysts can dive into the characteristics of each market to determine the influence of the timing of visitation and spend on the overall seasonal spread of tourism in the region.

Summary

The chapter has provided an introduction and context to the relevance and implications of temporality in tourism. It first established a framework through which the temporal lens of tourism activity can be appreciated. It then explored constructs of 'seasonality' and 'periodicity' whereby the former is understood to signify the peaks and troughs in demand during measurable time periods (typically monthly or quarterly) and resonates at destination level while the latter signifies imbalances in shorter operational timeframes for a business (within a day, between days, within a week, for example).

A total reliance on demand characteristics and patterns to explain temporal imbalances can be considered as an oversimplification, since the temporal dynamic of tourism involves supply-side considerations. Some of these (staffing and operational resource constraints, small business motivations, carrying capacity limitations) form the subject of ensuing chapters. Moreover, a range of underlying causal factors over which destinations or businesses have limited or no control also have a large part to play in understanding the temporal dimension of tourism. These are explored in the following chapter.

Self-assessment questions

1. Why do destination managers and tourism agencies (such as the UNWTO) collect data on the seasonal spread of tourism?
2. What are the distinctions between 'periodic' variation and 'seasonal' variations in tourism?
3. Why do tourism data collection agencies need an understanding of both short-term and longer term temporal variations in tourism activity?
4. Using UNWTO or Eurostat data for a few countries, describe their seasonality profiles – either in relation to the terms identified in Table 1.2, or your own terminology.

References

Bar On, R. (1975). *Seasonality in Tourism: a Guide to the Analysis of Seasonality and Trends for Policy Making*. Economist Intelligence Unit: London.

Baum, T. & Hagen, L. (1999). Responses to Seasonality: the experiences of peripheral destinations. *International Journal of Tourism Research* (1), 299-312.

Beaver, A. (2005) *A Dictionary of Travel and Tourism Terminologies* (2nd edn.). Wallingford UK: CABI Publishing.

Butler, R.W. (2001). Seasonality in Tourism: Issues and Implications. In Baum, T. and Lundtorp, S. (eds). *Seasonality in Tourism* Pergamon: Oxford. pp 5-21.

Butler, R.W. (1994). Seasonality in Tourism: issues and problems. In Seaton AV et a.l (eds) *Tourism: the State of the Art*, (Chapter 34). Wiley, London.

Boyer, M. (1972). *Le Tourisme*. Paris: Editions du Seuil.

Butler, R.W. & Mao, B. (1997). Seasonality in Tourism: Problems and Measurement, in Murphy, P. (ed.) *Quality Management in Urban Tourism*. Chichester: John Wiley and Sons. pp 9-23.

Chen, T. & Pearce, P. (2012). Research note: Seasonality patterns in Asian tourism. *Tourism Economics*, 18 (**5**), 1105-1115.

Connell, J. and Page, S. (2015). Visitor attractions and events. In Page, S. (ed.). *Tourism Management*, 5th ed. Routledge: London. pp 271-303.

Duro, J.A. & Turrión-Prats, J. (2019). Tourism seasonality worldwide. *Tourism Management Perspectives*, 31, 38-53.

Eurostat (2021) *Seasonality in tourism demand*. https://ec.europa.eu/eurostat/statistics-explained/index.php?title=Seasonality_in_tourism_demand.

Frechtling, D.C. (2001). *Forecasting Tourism Demand: Methods and strategies*. Butterworth- Heinemann.

Getz, D., Carlsen, J. & Morrison, A. (2004). *The Family Business in Tourism and Hospitality*. CAB International: Wallingford, UK.

Getz, D. & Nilsson, P.A. (2004). Responses of family businesses to extreme seasonality in demand: the case of Bornholm, Denmark. *Tourism Management*, 25, 17-30.

Goulding, P. (2006). Conceptualising supply-side seasonality in tourism: A study of the temporal trading behaviours in Scotland. Unpublished PhD, University of Strathclyde.

Goulding, P. & Pomfret, G. (2022). Managing temporal variation at visitor attractions. In Fyall, A., Garrod, B., Leask, A. & Wanhill, S. *Managing Visitor Attractions* (3rd ed.) pp 213-233. Routledge: Abingdon.

Grant, M., Human, B. & Le Pelley, B. (1997). Seasonality, *Insights*. London: BTA/ETB. July ppA5-A9.

Hartmann, R. (1986). Tourism, seasonality and social change. *Leisure Studies* 5(1), 25-33.

Haulot, A. (1963). L'Établissement des saisons touristiques. Nouvelle contribution à la solution d'un vieux problème, *Etudes et Mémoires*, Centre d'Etudes de Tourisme: Aix en Provence. pp 339-360.

Leiper, N. (1990). *Tourism Systems: An Interdisciplinary perspective*. Palmerston North, NZ: Massey University.

Lundtorp, S. (2001). Measuring tourism seasonality. In Baum, T. and Lundtorp, S. (eds.) *Seasonality in Tourism*, Pergamon: Oxford. pp23-50.

Page, S.J., Brunt, P., Busby, G. & Connell, J. (2001). *Tourism: a Modern Synthesis*, London: Thomson.

Su, M.M. & Wall, G. (2016). A comparison of tourists' and residents' uses of the Temple of Heaven World Heritage Site, China. *Asia Pacific Journal of Tourism Research*, 21 (**8**) pp905-930.

UNWTO (2021) *Global and Regional Tourism Performance: Seasonality 2019* https://www.unwto.org/tourism-data/global-and-regional-tourism-performance Accessed 5th June 2022.

Welcome to Yorkshire (2020). *Tourism Data Report*, March 2020. www.yorkshire.com, Accessed 21st March 2021 via www.visitbritain.org.

2 Causal Factors of Seasonality and Temporal Imbalances in Tourism

Philip Goulding

Learning outcomes

Having read this chapter, students should be able to:

1. Assess the relevance of a broad range of factors that influence the causation of temporal imbalances in tourism.
2. Identify, explain and apply a range of institutional factors that influence temporal demand patterns.
3. Distinguish how demand-derived and supply-side factors are moderated by interventions from various types of intermediaries.
4. Appreciate the role of business operating behaviours and motivations in affecting temporal patterns, particularly at local levels.

Introduction

The previous chapter established the prevalence of temporality in tourism, discussing dimensions, characteristics and terminologies of the phenomenon. It identified the traditional label of 'seasonality' as the main characterisation of temporal imbalances, while introducing 'cyclical' and 'periodic' dimensions, the latter more significant at micro-level, such as the availability or non-availability of tourism services within a shorter timeframe. The focus of this chapter is to unravel underlying causes of temporal fluctuations in tourism, to help build an understanding of why temporal peaks and troughs in demand patterns exist, what factors affect the temporal nature of consumers' holiday and wider travel patterns and therefore in turn, influence service providers in terms of the availability, nature and pricing of their tourism services.

A 'causal influence framework' is introduced to provide an overview of the various elements and their inter-relationships, after which each is explored in turn. However, while climate is a fundamental causal factor in explaining the temporal nature of tourism, it is examined more fully in Chapters 3 and 4. Hence the various perspectives of climate as a causal factor of temporal imbalances in tourism are not discussed in detail within the current chapter.

An overview of causal factors for temporality in tourism

Tourism is predicated by time, often characterised by waves of movements within and between countries that vary according to time. Bar On's (1975) pioneering longitudinal study of seasonal variations in 16 countries over a 10 year period not only provided ample evidence of the universality of seasonality in tourism but also highlighted the commonality of causal factors. Butler (2001) suggested that the pattern of tourism seasonality at any specific destination will be the result of interactions between a number of causal elements in both the generating (origin) and the receiving areas (destinations) and further modified by distribution channels and transport availability and by intermediaries, which may include tourism and non-tourism entities spanning the public/governmental and private (corporate/small business) sectors or a coalition of these. Figure 2.1 provides an adaptation and update of Butler's original model.

The inference of the framework is that no single factor is entirely responsible for the seasonal or temporal patterns of tourism in any specific location, rather, it is explained by the interplay of a number of factors across the spectrum of demand, supply and the moderating forces of the business environment and external interventions. They are not mutually exclusive. The chapter goes on to discuss and exemplify some of these causal factors.

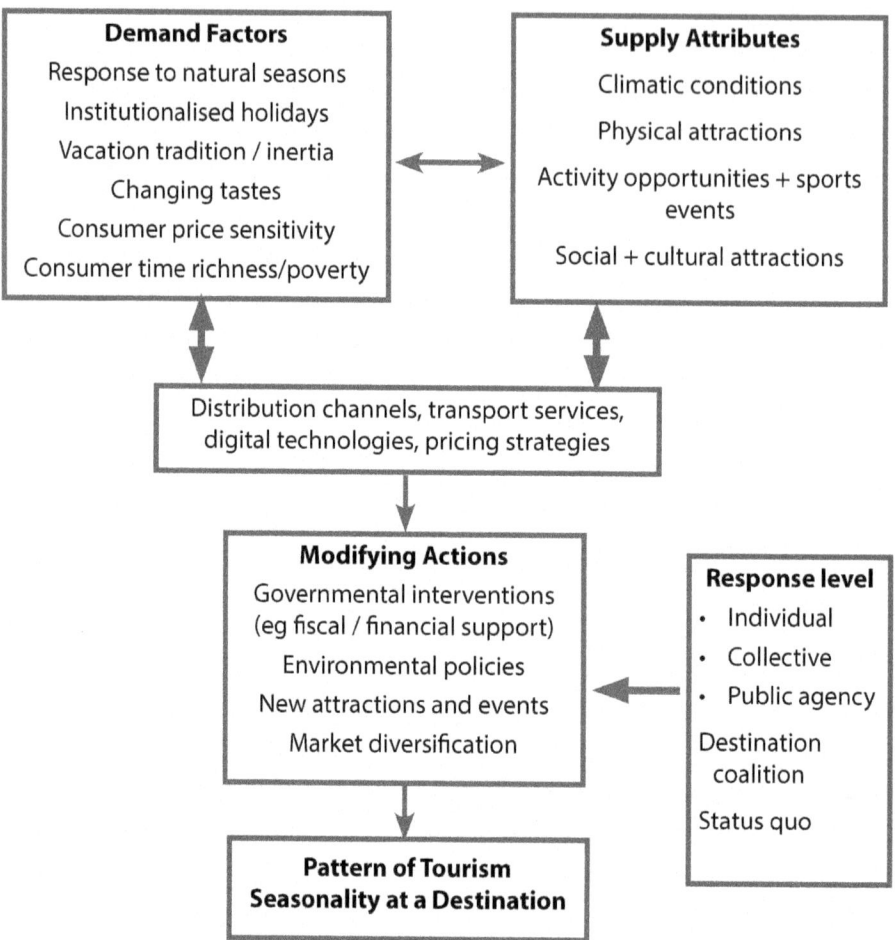

Figure 2.1: A framework of causal influences on tourism seasonality. Source: Adapted and updated from Bar On (1975), Butler (2001) and Goulding (2006)

Demand-derived causal factors

In the above model, demand factors include the push or pull of natural seasons as a stimulant to travel. They may trigger an 'escape' (a push factor) from prevailing adverse weather in the home environment (such as an extended cold weather patch, rain, damp in winter or even the prevalence of a heat wave in summer). Conversely, as a pull factor, the climatic characteristics of natural seasons have for centuries provided the 'pull' attraction of certain destination regions. For example, in Chapter 5, Crous-Costa et al. assess the role of climate in the development of health tourism in France and Switzerland in the 18th and 19th centuries.

Other demand causation factors reflect behavioural, economic and personal circumstances of tourism consumers (Bar On 1975; Frechtling 2001). Apart from disposable income as an influence or limitation on travel choices, 'time poverty' for some consumers may also dictate their temporal travel patterns in ways that reinforce mainstream tourism flows. For example, time availability may determine when people can and cannot take holidays, especially if self-employed or constrained by limited annual leave entitlements. 'Time rich' consumers, whether retired, working flexibly or for whom work and leisure boundaries blur, may similarly choose to seek out vacation periods that afford them escape from crowds. Likewise, Hartmann (1986) noted that 'vacation tradition' or inertia in the timing of holidays can be a potent force in stimulating shoulder seasons (the 'we always go away in late September to avoid the summer crowds' syndrome), even when the original time constraints (such as being bound by academic terms, when children have grown up, retirement etc), have disappeared.

Finally, Butler (2001) added *'social seasonality'* as a temporal pattern influence, referring to the historically 'rooted in lifestyle' leisure patterns of the aristocracy and elite classes, though gradually 'democratised' through time. He adds that 'social seasonality' now includes well defined seasons for attendance at activities such as classical music festivals, race meetings, regattas and so forth.

In addition to the above causal factors, the impact of the Coronavirus pandemic on previously established work patterns, along with the complexities of emerging work and lifestyle patterns in tourism generating societies, will require a re-categorisation of seasonal causation in future years. As an example, in the UK a government supported trial of a four day working week by 70 companies over six months is in its early stages (Kollewe, 2022). The effects of this on recreational travel patterns will be eagerly awaited by the tourism industry.

Institutional factors

While it is evident that temporal variations in tourism are essentially a demand-led phenomenon, based on consumers' travel motivations, economic and behavioural characteristics, according to Bar On (1975), Hartmann (1986) and Butler (2001), tourism flow patterns are also influenced by 'institutional factors' within generating areas. Institutional factors, from the tourism demand perspective, are essentially facilitators, embracing a range of cultural, statutory and religious celebrations which translate into opportunities afforded to people to travel during certain defined time periods. They may also be given 'public holiday' status by national or regional governments and therefore of statutory status. Hence the term 'institutionalised holidays', although this is a misnomer in some cases. Higham and Hinch (2002) noted that at the heart of institutional factors is a 'work vs leisure' perspective of time that, in Western societies, stems back to the Industrial Revolution. This resonates in societies where there is a prevalence of statutory public holidays including national days, labour days, bank holidays and so on, around which people can build 'long weekends' for leisure breaks. However, Bar On (1975) cast the 'institutional' net wider, including:

- Holidays and other events at specific fixed times of the year, such as Christmas and the summer vacations of schools and, as a legacy in some societies and industries, in places of work. While the practice of fixed time-specified holiday entitlements for workforces has declined significantly in recent decades in line with changing work patterns, there nevertheless remains a cultural legacy in some societies (Japan, South Korea), in which fixed holiday periods and work 'shut-downs' still exist. In Japan, for example, many manufacturers close factories during 'Golden Week', which is a string of Japanese public holidays that runs from late April till early May (Tanabe, 2022). As such, Golden Week results in a surge in travel movements within Japan and beyond its borders.

- The fiscal and financial trading year, affecting the budgets and taxes of individuals and businesses. In an empirical study of small tourism and hospitality business owner-operators in Scotland, Goulding (2006) found that the operators sometimes ceased or curtailed trading in the weeks approaching the end of the tax year if their business turnover was approaching the threshold for paying VAT.

- Variable date festivals (religious and non-religious), the temporal effects of which vary from year to year depending which week(s) and month the festival falls. The Lunar New Year in East and South-East Asia is the largest such annual event globally and heralds one of the largest human travel movements on the planet each year. Its start is determined by the lunar cycle, the second new moon after the winter solstice, hence generally the holiday period varies between the 21st January and 20th February.

Table 2.1 provides examples of some of the world's largest human movements which represent 'social seasonality' with an element of statutory holiday entitlement (in the case of the Chinese New Year and 'festive season' around Christmas) and religious celebrations or pilgrimages (the Hajj, Kumbh Mela) which are not primarily 'public holidays' per se.

Supply-side attributes and modifying actions as causal factors

Temporal supply attributes (i.e. at the destination) include climatic conditions, the touristic, physical and natural resources of the destination area (i.e. infrastructures and superstructures) and the activities and events/festival calendars taking place in the destination. In the latter case, Table 2.1 illustrates the cross-over between institutional factors and the supply-side, where 'socially seasonal' travel patterns and institutionalised holidays, celebrations or purely religious motivations are facilitated by the prevalence of tourist infrastructures in the destinations to host them.

An example here is the phenomenon of the 'recreational sporting season' (Butler, 2001; Higham & Hinch, 2002) which may embrace active or passive recreation. The development of sporting seasons can be linked to specific climatic and natural resource requirements and reflect changing patterns of recreation and active lifestyle in many tourism generating societies. Since the mid-19th century, snow- and water-based recreations have developed into major pastimes, around resource condition linked seasons. In terms of passive recreation, football, rugby and tennis and F1 motor racing are built on distinct calendars, which represent national or international 'sporting pilgrimages' of fans or supporters.

As previously noted, climate and weather are strong influences on visitor movements globally and regionally. They act as both pull and push influences on visitation patterns, determining demand at certain times, as discussed above. Numerous types of extreme weather events are seasonally defined, such as hurricanes and tropical cyclones, heat waves and flooding. They exert strong influence on travel patterns to the Caribbean, southern USA, parts of east Asia and the Indian Ocean region, through service closures and travel disruptions (Amelung, et al., 2007). Accelerating changes to climatic patterns are, however, proving challenging to international tourism as the frequency and timing of previously seasonally defined extreme events becomes ever less predictable.

Table 2.1: Examples of 'institutional' factors that significantly affect temporal tourist movements. Sources: Statista; Britannica; The National (India)

Primary locations	Fixed or variable dates	Duration	Status
Lunar New Year (Chinese Spring Festival, Vietnamese Tết, Korean Seollal). Commonly called 'Chinese New Year'			
East & South East Asia + Chinese, Korean, Vietnamese diaspora	Variable: lunar cycle, between 21st January – 20th February	Chinese Spring Festival 15-16 days, including Lantern Festival. Typically 3 days for Tết and Seollal and lunar holiday in other East/SE Asian countries	Partly official public holiday. Partly non-statutory cultural/social vacation period. Basis of a 'holiday season' and generates very large-scale travel movements, (c415m in 2020)
Christmas / 'festive season'			
Christian countries	Christmas: Fixed, 25th Dec. in Catholic and Protestant Gregorian calendars, or 7th Jan. in Orthodox countries (Julian calendar). Festive season: USA from Thanksgiving (late Nov.) through early Jan.	Typically 2 days public holidays.	Religious and cultural celebration + public holiday. Basis of a 'holiday season' sometimes extended to include Gregorian New Year period (early January). Generates very large-scale holiday and VFR travel movements.(c 115m in USA 2020 (Statista).
Hajj			
Islamic countries + diaspora	Variable: based on lunar calendar.	c1 month during the last month of the Islamic calendar.	Religious pilgrimage. Not a statutory holiday. Generates large scale travel movements (c2.5m in 2020). to/from Mecca.
Hindu festivals: Holi; Diwali/Deepavali; Kumbh Mela			
Hindu, Sikh, Jain and other religions in India + diaspora (eg UK, USA, Fiji, Mauritius). Hindu festival + pilgrimage.	Variable: based on lunar/solar calendar. Holi : Spring; Diwali : autumn Kumbh Mela : every 3rd year in a 12 year cycle at different times and locations in India.	Holi: 1-2 days ; Diwali: c2-5 days in autumn; Kumbh Mela: several weeks spread over the year	Religious celebrations Generate large scale leisure travel movements, esp. Kumbh Mela : c 240m in 2019 inc 1m overseas visitors

Modifying actions typically include the pricing strategies and tactics of travel and tourism service operators but may also include government interventions. Fiscal incentives or disincentives for travelling at certain times of the year, such as tourist taxes or environmental taxes levied by governments on destination services at certain times of the year are an example of this. The governments of Bhutan, Croatia and the regional government of the Balearic Islands all impose levies on visitors adjusted according to the season.

In 2020, the Coronavirus pandemic was a major cause of temporal variation, through government sanctioned suspension or reduced schedules of transport services as well as the closure of destination services such as visitor attractions, at various times during the year in many places across the world. This signalled government enforced temporal restrictions on business operations and on travelling, as an extreme 'moderating' factor.

Meanwhile Baum and Hagen (1999), Flognfeldt (2001) and others have demonstrated the role of labour as a constraint to the operation of a year-round tourism season, an issue illustrated by Baum et al. in Chapter 10. In extreme cases, local labour shortages in the visitor service sectors, without the ability of local areas to attract in-bound seasonal labour, can impact on the functioning of the local tourism economy. This phenomenon has been widely illustrated in empirical studies in the Danish island of Bornholm (Getz & Nilsson, 2004), in Crete (Kousis, 1989) and in the Carolinas in the USA (Terry, 2016).

Local stakeholders can also intervene in the operation of the tourism market, in particular where there are problems arising from excess demand in peak seasons. Environmental constraints and visitor management policies may be implemented with the aim of spreading demand and resource use more evenly across the year. Local consortia may also use market diversification as a tool, in an aim to promote the destination to consumer segments who are more time flexible and thus more likely to travel in shoulder and low seasons. Another source of modifying actions are regulations and controls from local government that may also restrict year-round trading or define a 'season'. In the UK, such is the case for the operating season of some holiday home parks, for fishing rights on many rivers and weather-dependent health and safety regulations on activity operators' duty of care to customers. Accordingly, modifying actions by intermediaries, governments and/or suppliers may influence the seasonal pattern of tourism through restricting capacity or supply at certain periods, even temporarily.

The final key component of supply which deserves attention are the local business operators themselves. As will be demonstrated by Holland in Chapter 11, trading decisions by tourism owner-operators and self-employed service providers who are essentially 'lifestyle' operators can affect the supply and hence capacity and can influence demand patterns in a locality to host tourism at certain times of the year. This is especially so in rural locations.

Figure 2.2: Temporal closure: when the proprietors take vacations. Puerto Vallarta, Mexico. Image: the author.

Vignette: 112 Reasons to trade seasonally

In a large scale empirical study of small and largely independent tourism related businesses in Scotland, Goulding (2006) found that seasonal trading was widespread throughout the country, but especially so in the more rural and remoter areas in the south and in the Highlands and Islands. 'Seasonal trading' ranged from temporary closures during public holidays (typically Christmas and New Year) and at other times of the year for typically a week at a time, to several months' duration. The extended closures were particularly prevalent over the winter months. Hence, there was often a distinct 'operating season' and a 'closed season' for a plethora of farmhouse accommodation operators, private apartments, for local private museums and galleries and other attractions, cafes, small restaurants and other eateries and for commercial home enterprises such as bed and breakfasts and guest houses (Goulding, 2009). At the other end of the temporal spectrum, some operators closed the premises temporarily at short notice in response to unforeseen factors such as a family crisis or health issue or a technical issue that compromised the service. Staffing cover was also a factor, if the operation relied on a key staff member (for example a chef) who called in sick.

Table 2.2: Motivational and Influence variables on temporal trading by tourism business proprietors. Source: Goulding (2006)

Variable Type			
Economic	Exogenous	Intrinsic Personal	Market	Natural
Variable Clusters				
viability of staying open. operational costs: overheads, utilities, maintenance, upgrading. fiscal: community charge, staying below the VAT threshold. staffing: labour uncertainties yield: return on capital, turnover. work related: hours of work, workload, other work commitments.	institutional: the school year; calendar effects. role of public agencies: financial assistance, licences, bureaucracy. location related: location, site closure, cost of access to the destination area, access to property. transport related: transport infrastructure, frequency. destination issues: other local amenities closed, lack of destination facilities. Public health: epidemics (on farm)	work-life balance: free time, rest/relaxation, escape/get away, lifestyle specified. social priorities: friends visiting, family commitments, community networks. internalised variables: privacy, self-occupancy (of rented apartments). lifecycle and health: age, retirement, energy levels, mental wellbeing. migration altruism: conservation, environment, local community benefits, local economic impacts.	state of the market: lack of business, type/nature of tourism, competition, tourist perceptions, market trends. product related: season-specific recreations (golf/fishing). flora/fauna/wildlife resources. business configuration: multi-service operation. student lets. marketing responses: low take-up of marketing campaign. Business responses: developing new services/products	climate/weather. natural seasonal conditions: daylight hours, tides. individual climatic variables: frost, extreme cold, rainfall, snow, wind. Weather consequences: dampness, flooding, customer safety.

> The research question to the proprietors sought to uncover the motivators and influences behind the trading decisions. In other words, the question 'why don't you operate the business on a full year-round basis?'. From 1,292 responses, 112 reasons emerged which could be grouped into five distinct variable categories. Table 2.2 captures a selection of those variables.
>
> Probed on their attitudes to seasonal trading, the findings from the traders indicated a marked preference for the operating season or period to fit with proprietors' lifestyle choices and an overall preference for seasonal vs year-round trading. Yet for a majority, maximising revenues was considered important.
>
> Overall, the study demonstrated that, from a supply-side perspective and at individual and largely small-scale business levels, temporality was an inherent characteristic of Scottish tourism beyond the large conurbations.

Summary

In this chapter, to make sense of the many disparate factors contributing to the causation of seasonality or temporal imbalances in tourism, we have employed and built upon a conceptual framework developed by Butler (2001). We have seen that the causal framework incorporates distinct demand-derived and supply-side factors, modified by the interventions of intermediaries, including industry, government and local stakeholder coalitions. It has been illustrated that the framework is fluid, in that there is significant overlap and inter-relationship between the various factors. For example, institutional factors may on the one hand represent stimuli, constraints or facilitators to travel demand. On the other hand, the statutory nature of some of them are more akin to interventions or modifying actions.

What is clear, however, is that tourism's seasonal and temporal patterns are dynamic and fast evolving. The Coronavirus pandemic has led to significant shifts in work-life patterns which will in turn impact over time on the temporal patterns of human movements. Climate change will also, over time, redefine temporal movements in response to weather events and global warming patterns. Add to this the rapid development of digital technologies that enable travel decision-making and it is evident that the quest to characterise and understand the causes of temporal imbalance will become even more complex.

Self-assessment questions

1. Consider the framework of causal influences for seasonality and temporal imbalances in tourism (Figure 2.1). What additional demand-related factors would you add to the framework and why?
2. How might the temporary or seasonal closures of tourism/hospitality related businesses affect tourism in their localities?
3. Consider the major human movements outlined in Table 2.1. What others can you think of that impact on the seasonal spread of tourism in your destination country or locality? For example, public holidays, cultural celebrations, particular events and so on?
4. Thinking of your peer group or family, what factors most influence the time of year and the duration of travel they undertake? Where would these factors be categorised in the framework (Figure 2.1)?

References

Amelung, B., Nicholls, S. & Viner, D. (2007). Implications of global climate change for tourism flows and seasonality. *Journal of Travel Research*, 45, 285-296.

Bar On, R.R. (1975). *Seasonality in Tourism: A Guide to the Analysis of Seasonality and Trends for Policymaking*. London: Economist intelligence Unit, Technical Series No. 2.

Baum, T.G. & Hagen, L. (1999). Responses to Seasonality: the experiences of peripheral destinations. *International Journal of Tourism Research*, 1, 299-312.

Butler, R.W. (2001). Seasonality in Tourism: Issues and Implications. In T.G Baum and S. Lundtorp (eds.), *Seasonality in Tourism*. Amsterdam: Pergamon.

Flognfeldt, T. (2001). Consequences of Summer Tourism in the Jotunheimen Area, Norway. In T. Baum & S. Lundtorp (Eds.), *Seasonality in Tourism* (pp 109-118). Amsterdam: Pergamon.

Frechtling, D. (2001). *Forecasting Tourism Demand: Methods and strategies*. Oxford: Butterworth-Heinemann.

Getz, D. & Nilsson, P.A. (2004). Responses of family businesses to extreme seasonality in demand: the case of Bornholm, Denmark, *Tourism Management*, 25(1), 17-30.

Goulding, P. (2006). 112 Reasons to trade seasonally: A motivation and influence paradigm of Scottish seasonal tourism businesses. *Proceedings of the Third Graduate Research in Tourism Conference, May 2006*. Anatolia International Journal of Tourism and Hospitality Research and Çanakkale Onsekiz Mart University. Çanakkale, Türkiye.

Goulding, P. (2009). Time to Trade? Perspectives of temporality in the commercial home enterprise. In P. Lynch, A.J. McIntosh & Tucker, H. (eds.), *Commercial Homes in Tourism: An international perspective*: Abingdon: Routledge.

Hartmann, R. (1986). Tourism, seasonality and social change, *Leisure Studies*, 5(1), 25-33.

Higham, J. & Hinch, T. (2002). Tourism, sport and seasons: the challenges and potential of overcoming seasonality in the sport and tourism sectors, *Tourism Management* 23, 175-185.

Kollewe, J., (2022). Thousands of UK workers begin world's biggest trial of four-day week. *The Guardian*, 6 June. www.theguardian.com/business/2022/jun/06/thousands-workers-worlds-biggest-trial-four-day-week. (Accessed 8 June 2022).

Kousis, M. (1989). Tourism and the family in a rural Cretan community. *Annals of Tourism Research,* 16 (3), 318-332.

Tanabe, S. (2022). Toyota, Subaru extend Japan holiday shutdowns over supply crunch, *Nikkei Asia.com*, 22 April. https://asia.nikkei.com/Business/Automobiles/Toyota-Subaru-extend-Japan-holiday-shutdowns-over-supply-crunch (Accessed 15 June 2022).

Terry, W.C. (2016). Solving seasonality in tourism: Labour shortages and guest worker programmes in the USA. *Area*, 48(1), 111-118.

3 Tourism and the Seasons

Adele Doran and Peter Schofield

Learning outcomes

This chapter will provide you with:

1. An understanding of how climate and weather influence global seasonal tourism demand.
2. An appreciation of how institutional seasonality exacerbates the effects of natural seasonality.
3. An awareness of how seasonal tourism demand will alter as a result of climate change.

Introduction

This chapter focuses on climate as a key causal factor and determinant of seasonality. It will explore our understanding of how climate acts as an important construct to patterns of tourism in various parts of the world. It will illustrate how the weather in both the generating regions and the destination areas produces 'push' and 'pull' factors influencing tourism demand. It will also assess how institutional seasonality exacerbates the effects of natural seasonality creating peaks in tourism demand. Finally, the chapter will examine how global warming is changing the seasons, redistributing climatic assets among tourism regions and influencing global tourism demand.

Climatic variation and tourism demand

Weather is the atmospheric conditions over a short period of time and is affected by a number of factors including temperature, humidity, cloud cover, wind and precipitation. Climate is the weather averaged over a long period of time and represents the conditions anticipated at a specific destination and time. Climate defines the length and quality of tourism seasons in leisure destinations and determines a destination's attractiveness, such as the temperature, snow conditions, and wildlife productivity and biodiversity. Therefore, it is a principal driver of global seasonality in tourism demand (Mintel, 2012; UNWTO, 2008).

Climatic seasonality represents a significant challenge for the tourism sector due to the uneven nature of demand for visitor attractions and accommodation, and the relatively fixed nature of the supply of capacity and resources (Hadwen et al., 2011). Seasonality is driven by the permanent 22.5 degree tilt of the earth's axis as it orbits around the sun, which means that throughout the year different parts of the planet's surface are exposed to direct solar rays which impact the environment and human behaviour (Ulijaszek & Strickland, 2009). It is summer in the northern hemisphere when it's tilted towards the sun, winter in the southern hemisphere when it leans away from the sun and vice-versa. Correspondingly, climatic seasonality is less marked at the equator compared with higher latitudes, both north and south. The variation in sunlight at different times of the year influences global wind patterns, ocean currents and atmospheric moisture levels, which also contribute to seasonality together with the earth's topography, particularly altitude. While this natural seasonality has been considered to be relatively permanent and predictable, there is increasing evidence that the seasons are changing because of global warming, which is discussed later in the chapter.

Crucially for tourism, the weather changes in each season, in both the generating regions and the destination areas and this produces push and pull factors influencing tourism demand (Figure 3.1).

Climatic seasonality and its attendant differences in weather between tourists' origin areas and their intended holiday destinations is a key influence on tourists' decision making. In addition to this natural seasonality, 'institutional seasonality' (Butler, 1991), as discussed in the previous chapter, a combination of religious, social and cultural factors, also affects demand.

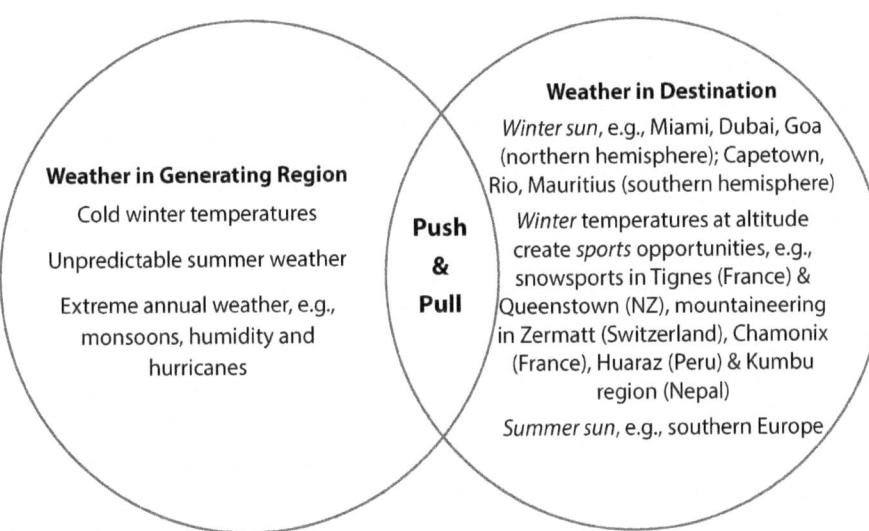

Figure 3.1: Climatic push and pull factors in tourism

Institutional causes include the timing of school and work holidays, individual preference for the traditional peak season, whether or not destination resorts are in fashion, and the programming of festivals and events at tourist destinations. For example, while tourist visitation to areas in eastern Australia was found to be driven primarily by climatic seasonality (Hadwen et al., 2011), Mediterranean destinations, e.g. Sicily, despite having a favourable all year-round climate, experience a single summer peak in demand because of the influence of institutional seasonality (Cuccia & Rizzo, 2011). By comparison, some destinations e.g., Uganda experience 'two peak' seasonality. Uganda's tourism peaks in July/August and December/January reflect both weather patterns and the summer and Christmas holidays, respectively in its key European and North American markets. Therefore, although tourism is a climate-dependent industry, many destinations owe their popularity to their agreeable climates during traditional holiday seasons, but this compatibility of climatic and institutional seasonality is likely to be challenged by global warming and its impact on climate change.

Seasonal variation and visitation

Seasonal variation in tourism visitation is influenced by a number of factors including the attributes of a destination, the characteristics of the generating region and the specific markets being targeted. As a result, destinations in the same region, e.g., Europe, can experience different degrees of seasonality because of competition for both intra- and extra-European visitors. From an empirical perspective, seasonal patterns of tourism have been classified in a number of ways from destinations with a 'single peak' summer season typical of Mediterranean coastal areas, to destinations with a 'peak and shoulder season' (or minor peak before the off season), through to 'two peak' destinations such as mountain resorts with summer and winter seasons (López-Bonilla et al., 2006). In addition, there are also 'no peak' destinations with low tourism seasonality e.g., cultural city destinations. To date, there have been few large-scale geographical analyses of tourism seasonality (Coshall et al., 2015); however, some examples of seasonal variations and tourism visitation in tropical, temperate and sub-arctic locations will help to illustrate the impact of climatic and institutional seasonality on tourism patterns, notwithstanding the regional and national contexts (Table 3.1).

Table 3.1: Seasonal variations and tourism visitation in tropical, temperature and sub-arctic locations

Location	Climate	Major Destinations
Tropical areas		
Located between 23.5° north and south of the Equator	Receives more direct sunlight than the rest of the planet Hot and humid 'tropical' climate Temperatures of at least 18°C High annual rainfall Little climatic variability	Queensland, Australia; Kauai, Hawaii; Seychelles; Cook Islands; South East Asia, inc. Vietnam, Cambodia, Thailand, Malaysia, Bali
Temperate areas		
Located between 35° and 50° north and south of the Equator	Four distinct seasons Warm summers, cold winters and moderate shoulder seasons Mean temp. −3°C to 18 °C Two climates: maritime (cool summers and mild winters) continental (hot summers and very cold winters)	West coast of North America, Western Europe and southern parts of Australia and New Zealand (Maritime) Central parts and east coast of North America, Eastern Europe and Asia (Continental)
Sub-arctic areas		
Located between 50° and 70° north of the Equator	Some of the most extreme seasonal temperature variations Long winters with temperatures below −50°C in extreme cases Summers are warm, but short (max. 3 months) with temperatures sometimes reaching 26 °C	Alaska, Canada, Siberia, northern Scandinavia, northern Scotland, Iceland and the Shetland Islands

Tropical areas

Even at this latitude, with less climatic variation than in temperate areas to the north and south, managing tourism seasonality is still a concern. For example, Singapore must consider institutional seasonality and variation in expenditure between national visitor segments, in addition to climatic seasonality in its key market areas. Japan and China represent the main geographic markets. China is the largest market, but Japanese tourists' expenditure is higher by comparison (Hui & Yuen, 2002). Visitation from Japanese tourists consistently peaks in July/August and December, but while it would seem preferable to stimulate demand in the spring and autumn, Japan's pleasant weather and absence of school holidays at these times constrain this option. Therefore, despite the over-demand for facilities and amenities at the peak times, Singapore prefers to attract more Japanese tourists or encourage them to extend their stay.

Figure 3.2: Phu Quoc Island, Cambodia: a tropical tourism destination.
Author's image

Temperate areas

While climatic seasonality can be a key issue affecting visitation to destinations in the temperate zone, these areas include many of the world's major tourism cities, e.g., New York, London, Paris, Moscow and Beijing – 'no peak' destinations – which attract leisure visitors in all seasons because of their distinctive cultural attractions (Rutty & Scott, 2010). Moreover, most of the world's population lives in this climatic zone and given the unpredictability of summer weather in many temperate areas and the fact that destinations with sunshine boast the highest tourism demand, the temperate climate is a major push factor in global tourism, including the largest single flow of tourists in the world from temperate northern Europe southwards to the Mediterranean's predictable sunny weather.

Figure 3.3: Morzine, France: a snow sports destination located in a temperate area. Author's image

Sub-arctic areas

These areas experience high climatic seasonality with respect to tourism flows and visitor activities, but tourism has grown significantly since the late 2000s, including winter cruise passengers. For example, Northern Norway, inspired by the success of Finnish Lapland, now offers northern lights tours and snow-based products to offset the effects of their extreme seasonality (Jaeger & Viken, 2014). However, prospective tourists' negative images of winter conditions in subarctic areas, and both intrapersonal and structural constraints, have continued to limit winter visitation despite recent changes in visitor perceptions relating to the aesthetic qualities of winter landscapes, romantic notions of snow and darkness, and both soft adventure and soft exploration tourism.

Case study: Winter sun - spring break in Miami

'Spring break' in Miami illustrates the interplay of climatic and institutional seasonality with respect to the appeal of the tropical climate for students on their spring break from university, which is typically taken in March and centres on the destination's beaches. While climatic preferences and thresholds for beach tourism vary according to visitor origin (Rutty & Scott, 2013), the allure of the weather in the 'Sunshine State' reigns supreme, particularly for those wishing to escape the sub-zero temperatures in northern US states. The country's top ranked spring break destination also offers free beachfront events, music festivals, nightclubs and bars along the 88 ocean front blocks of Miami Beach. Nevertheless, Bill Talbert (2015), CEO of Greater Miami Convention & Visitors Bureau's argues that *"destinations aren't about geography, they're about psychology...about a feeling"* This is exemplified by media induced impressions of Miami. Over and above its tangible assets, the city's cool, edgy image as portrayed in film and television, despite crime statistics to the contrary, is a significant pull factor for the spring break demographic. While South Florida's reputation as a drug-related crime capital did have a negative impact on tourism in the 1980s, the *Miami Vice* television series in 1984 transformed the city's identity. This stimulated inbound tourism, positive global brand awareness, the growth of Miami's advertising and fashion industries, and when Will Smith released his hit single, *Welcome to Miami* in 1998, the city region transitioned into mainstream tourism (Bohn, 2009). Currently, around 500,000 students take their spring break in Florida, a third of all U.S spring breakers. As such, this market makes an important contribution to Greater Miami's tourism industry and illustrates the impact of both climatic and institutional seasonality on the destination.

Impacts of annual and unusual weather patterns on tourism flows

Whilst good climate is a motivator for travel, expectations of poor weather and climate may constrain tourism to a destination. In particular, annual weather events such as cyclones, tornadoes and hurricanes can disrupt tourism activity in both the long-term and short-term. So too can sporadic and unusual weather patterns, such as El Niño (warm) and La Niña (cold), events which occur when the Pacific Ocean near the equator becomes significantly warmer than usual triggering or intensifying weather extremes across the world (WMO, 2016). Such extreme weather events, increasingly frequent in the first two decades of the 21st Century have included flooding in South America and East Africa, droughts in southern Africa, increased tropical cyclones in the Pacific, blizzards in the USA, cold and wet summers in Europe and drought in South East Asia and Australia.

Both annual and unusual weather events can impact the environment, tourism resources and infrastructure and they can alter tourists' perceptions of a destination's attractiveness and perceptions of risk and personal safety (Hall, 2018). For example, seasonal monsoons in the Nepalese Himalaya bring considerable rain lasting from a few hours to a few days, resulting in landslides, damaged roads and tourist trails, making it impossible for tourists to enjoy activities such as trekking, bird watching, elephant riding and sightseeing (Nyaupane & Chhetri, 2009). In addition, periodic water shortages during dry seasons can restrict water-based tourism activities. Accordingly, this restricts year-round tourism demand in Nepal. Conversely, whilst El Niño/La Niña can constrain tourist arrivals, when in the country of origin, they could also induce residents to travel to warmer countries. For example, La Niña in the USA acts as a push factor on tourist arrivals to the Philippines, rather than the tropical climate in the Philippines being a pull factor (Saverimuttu & Varua, 2014). Climate change will make such weather events more frequent and extreme, providing less time for physical and human systems to recover and it may result in long-term environmental deterioration (Hall, 2018).

Tourism is a highly climate-sensitive economic sector. A change in climate *"will alter seasonal tourism demand by creating, deteriorating or improving climatic conditions at destinations and in source markets"* (UNWTO, 2008, p.103). Tourists are flexible, and they will respond to climate change impact by substituting the place, timing and type of holiday, even at short notice (UNWTO, 2008). Therefore, it is expected that there will be a redistribution of climatic assets among tourism regions. Currently, projections of tourism demand as a result of a change in climate resources remain geographically limited to Europe (IPCC, 2018; Mintel, 2012; UNWTO, 2008). For example, winter sun destinations may see increasing competition from cities as cities become warmer in winter months. The Mediterranean is projected to become warmer and subject to more frequent heatwaves and tropical nights in the summer, making it less desirable at that time of year and more attractive in spring and autumn. Tourists from Northern Europe, who dominate international travel, are likely to spend more holidays in their own country or region as it becomes more suitable for tourist activities year-round. It is predicted that those who live in the warmer Mediterranean countries will also travel to these temperate countries in the summer. Winter sports destinations will continue to see a decline in natural snow and shortened ski seasons. Even with increased snowmaking, the ski industry is projected to contract with fewer operating ski areas, altered competitiveness among and within regional ski markets and a reduction in overnight stays. *"Consequently, there will be winners and losers at the business, destination and national level"* (UNWTO, 2008, p.61).

More flexible institutional holidays would help, as it would enable tourism demand to spread across a larger number of months and destinations (UNWTO, 2008). If institutional holidays remain the same, the geographic distribution of tourism is likely to be intensified in specific areas. For example, destinations like the Mediterranean resorts that will become too hot will not only see a decrease in visitation during the summer, but also during the shoulder seasons as institutional holidays will restrict people from visiting. However, an ageing population could increase demand in the shoulder seasons due to an increase of retirees and empty nesters who are not subject to the constraints of school holidays or professional responsibilities (Mintel, 2012).

In recent years, the global temperature has exceeded or been close to 1°C above the pre-industrial period (1850-1900) (IPCC, 2018). Global warming has already affected the environmental conditions which are a critical resource for tourism. Mountain, island and coastal destinations are particularly sensitive to climate change, as are destinations that are nature-based (Mintel, 2012; UNWTO, 2008). However, if the global temperature exceeds the Paris Agreement goal of 1.5°C and reaches 2°C, it may result in long-lasting and irreversible changes (IPCC, 2018). These include the erosion of beaches, coral bleaching, changes in snow cover, the loss of some ecosystems and reduced aesthetic appeal of landscapes (Mintel, 2012; UNWTO, 2008). Excess water due to extreme weather events, such as flooding and El Niño, will impact both natural and cultural heritage attractions, making these destinations less appealing to tourists. Furthermore, a 2°C warmer world would reduce European tourism by 5% with losses up to 11% for southern Europe (IPCC, 2018). Limiting global warming to 1.5°C would substantially reduce the probability of erratic wind and precipitation patterns, heavy flooding, extreme drought, and more intense heatwaves and tropical storms (typhoons and hurricanes). It will also reduce the retreat of glaciers and polar ice caps and a warming ocean surface temperature which are contributing to a rise in the sea level (IPCC, 2019).

These climate change impacts are leading to places being labelled as 'last chance' tourism destinations (Dawson et al., 2015). Consequently, 'last chance to see' tourism markets are developing where travellers visit these destinations before they are substantially degraded by climate change or to view the impacts of climate change on landscapes, seascapes, natural resources and/or social heritage (Dawson et al., 2015; IPCC, 2018); for example, to see vanishing glaciers, polar bears, historic sites and indigenous cultures. In turn, tourists are further deteriorating the destinations they are travelling to see and therefore accelerating their decline.

Case study: Climate change and ski tourism

Demand for ski-tourism is dependent on institutional seasonality (school holidays, half term and Easter), yet crucially, it relies on specific climatic conditions and is vulnerable to climate change. Natural snow depth and early snowfall have been found to have a positive impact on demand. Conversely, if snow conditions are poor, ski tourists will substitute spatially and visit another destination, substitute temporally and delay the trip or ski less often, or substitute for an alternative holiday activity (Steiger et al, 2019; UNWTO, 2008). Resorts above 2500m, where snow conditions are better, benefit from spatial redistribution of tourism demand (Mintel, 2019). Temporal substitution results in increased demand peaks, such as in January and February in the northern hemisphere when the snow is more reliable and alters the seasonal distribution of skier visits (Steiger et al., 2019). By comparison, activity substitution, which will reduce overall skier visits, is less common.

Recent seasons have been characterised by poor snowfall and extreme weather conditions which have contributed to declining participation (Mintel, 2019). Skiers have adapted by selecting ski areas that have greater snow-making capacities, invested in comfortable high-speed lifts and diversified their tourism products (Mintel, 2019; Steiger et al., 2019; UNWTO, 2008). This requires large capital investments and increased operating costs, forcing many smaller ski areas to close. Diversifying into four-season destinations increases and spreads tourism demand throughout the year and increases revenues (Steiger et al., 2019). For example, mountain biking on snow, winter music festivals, cricket on ice and indoor sky diving offer new activities for both winter and summer seasons (Mintel, 2019). These activities also speak to Millennials' and Gen Zs' desire for experiences. This is the world's biggest consumer group and whilst Baby Boomers and Gen Xers still dominate winter sports, they are an aging population. Finally, to mitigate snow condition concerns and to foster demand, ski resorts have introduced web cameras with real-time display of snow conditions on ski slopes and snow reports on social media (Steiger et al., 2019).

Conclusion

Weather and climate are the principal resources and constraints for global tourism demand patterns. They are considered consciously or implicitly throughout the tourists' planning process as significant motivators and important influences on destination choice, the timing of visitation and on overall visit satisfaction (Rutty & Scott, 2013). As such, the seasonality of tourist visitation to a given destination is strongly influenced by climate in both the origin and destination areas, but also by the complex interplay of other environmental, institutional, social and cultural variables together with tourism market and destination characteristics which together produce a range of push and pull factors. The relative importance of climatic and other factors, notably holiday periodicity, varies across the world according to climate zone and, in most zones, climate is the main factor influencing visitation seasonality.

Climate change is anticipated to have important consequences for tourism demand from the global to the destination scale. It will result in tourists substituting the place, time and type of holiday, therefore altering traditional seasonal destinations, such as the Mediterranean in the summer and mountain destinations in the winter. There is no evidence to suggest that climate change will reduce global tourism demand, instead the impact will occur at a destination level, as climate change shifts demand to other destinations that offer a more attractive climate (UNWTO, 2008). Consequently, given the prominence of climate, particularly temperature, for seasonal tourism patterns, advancing our understanding of tourists' climate needs regarding optimal and threshold visitation conditions and their variation by destination and visitor origin will be critically important for accurate forecasting of seasonality patterns as our global climate changes.

Self-reflection questions for students

1. How do climate and weather influence tourism demand?
2. Does seasonal variation in tourism demand always result from a combination of natural and institutional factors?
3. Identify a destination where climate change is influencing temporal demand, either positively or negatively. For example, reducing/lengthening the duration tourists stay in the destination or changing the time of year they visit. What is the destination doing to mitigate/encourage this change in demand?
4. Explain El Niño and La Niña and discuss the impact these events have on tourism, giving examples.
5. It is predicted that the Mediterranean will become seasonally 'too hot' for tourism within a few decades due to global warming. What are the implications for destinations in this part of Europe?
6. How can you personally make a contribution to reducing global warming and climate change?

References

Bohn, L. (2009) A brief history of spring break, *Time Magazine*, **173**(12), 14-15.

Butler, R.W (1991) Seasonality in tourism: Issues and implications, in T. Baum & S. Lundtorp (eds.), *Seasonality in Tourism, Advances in Tourism Research Series*, London: Routledge, pp. 5-18.

Coshall, J, Charlesworth, R. & Page, S. J. (2015) Seasonality of overseas tourism demand in Scotland: A regional analysis, *Regional Studies*, **49**(10), 1603-1620.

Cuccia, T. & Rizzo, I. (2011) Tourism seasonality in cultural destinations: Empirical evidence from Sicily, *Tourism Management*, **32**(3), 589-595.

Dawson, J., Lemelin, R., Stewart, E. & Taillon, J. (2015) Last chance tourism: A race to be last?, in M. Hughes, D. Weaver & C. Pforr (eds), *The Practice of Sustainable Tourism: Resolving the paradox*, Abingdon: Routledge, pp. 133-145.

Hadwen, W.L., Arthington, A.H., Boon, P.I., Taylor, B. & Fellows, C.S. (2011) Do climatic or institutional factors drive seasonal patterns of tourism visitation to protected areas across diverse climate zones in Eastern Australia?, *Tourism Geographies*, **13**(2), 187–208.

Hall, C. M. (2018) Climate change and its impacts on coastal tourism: Regional assessments, gaps and issues, in A. Jones and M. Phillips (eds), *Global Climate Change and Coastal Tourism: Recognizing Problems, Managing Solutions and Future Expectations,* Boston: CAB International, pp. 27–48.

Hui, T-K & Yuen, C.C. (2002) A study in the seasonal variation of Japanese tourist arrivals in Singapore, *Tourism Management,* **23**(2), 127–131.

Intergovernmental Panel on Climate Change (IPCC) (2018) Special Report: Global Warming of 1.5°C. Retrieved from: https://www.ipcc.ch/sr15/

Jæger, K. & Viken, A. (2014) Sled dog racing and tourism development in Finnmark, in A. Viken and B. Granås (eds.), *Tourism Destination Development: Turns and Tactics.* Farnham: Ashgate, pp. 131-152.

López Bonilla, J.M., López Bonilla, L.M. & Borja Sanz Altamira, B.S. (2006) Patterns of tourist seasonality in Spanish regions, *Tourism and Hospitality Planning & Development,* **3**(3), 241-256.

Mintel (2012) Tourism and Climate Change – International – June 2012. Retrieved from: https://reports.mintel.com/display/590197

Mintel (2019) Winter sports in Europe – September 2019. reports.mintel.com/display/924170

Nyaupane, G. P. & Chhetri, N. (2009) Vulnerability to climate change of nature-based tourism in the Nepalese Himalayas, *Tourism Geographies,* **11**(1), 95-119.

Rutty, M. & Scott, D. (2010) Will the Mediterranean become 'too hot' for tourism? A reassessment, *Tourism and Hospitality Planning & Development,* **7**, 267-281.

Rutty, M. & Scott, D. (2013) Differential climate preferences of international beach tourists, *Climate Research,* **57**, 256–269.

Saverimuttu, V. & Varua, M., (2014) Climate variability in the origin countries as a 'push' factor on tourist arrivals in the Philippines, *Asia Pacific Journal of Tourism Research,* **19**(7), 846-857.

Steiger, R., Scott, D., Abegg, B., Pons, M. & Aall, C. (2019) A critical review of climate change risk for ski tourism, *Current Issues in Tourism,* **22**(11), 1343-1379.

Talbert, B. (2015) https://skift.com/2015/09/10/interview-miami-tourism-ceo-explains-the-essence-of-destination-branding/ Accessed 15th October 2020.

Ulijaszek, S.J. & Strickland, S.S. (Eds.) (2009) *Seasonality and Human Ecology,* Cambridge: Cambridge University Press.

United National World Tourism Organisation (UNWTO) (2008) *Climate Change and Tourism – Responding to Global Challenges.* https://www.e-unwto.org/doi/book/10.18111/9789284412341

World Meteorological Organisation (WMO) (2016) *Exceptionally strong El Niño has passed its peak, but impacts continue.* public.wmo.int/en/media/press-release/exceptionally-strong-el-niño-has-passed-its-peak-impacts-continue

4 Nature and Time

Adele Doran and Seonyoung Kim

Learning outcomes

This chapter will provide you with:

1. A basic understanding of how nature-based tourism is defined and an awareness of related types of tourism.
2. Knowledge on the motivations of nature tourists and what is driving demand.
3. An appreciation of how the temporal dimensions of wildlife migration, valuing time in nature and vanishing natural resources, contribute to the attraction of nature-based tourism.

Introduction

Following on from the previous chapter's focus on climate and natural seasons, the purpose of this chapter is to explore the rich complexity of inter-relationships between nature and tourism, from temporal constructs. It will define nature-based tourism and explore what is driving demand. It will illustrate how temporal natural wildlife migrations and natural phenomena influence tourism demand. It will also assess how concern over vanishing natural resources is inducing some tourists to rush to visit before they are gone or irreversibly changed. Finally, it will examine the relationship between nature-based tourism and the temporal practices of slow tourism.

Defining nature tourism

Nature-based tourism is tourism centred on the natural attractions or resources of an area. The term is closely associated with other types of tourism, such as ecotourism, adventure tourism, wildlife tourism, and wilderness tourism (see Table 4.1). Although these types of tourism often take place in the same environment (e.g., national parks, nature reserves, protected areas) and the terms are often used interchangeably, it should be noted that there is an ongoing debate about the definitions of and relations between these different types of tourism.

Table 4.1: Defining nature-based tourism and other related forms of tourism

Typology	Definition	Example activities
Ecotourism	"Responsible travel to natural areas that conserves the environment, sustains the well-being of the local people, and creates knowledge and understanding through interpretation and education of all involved: visitors, staff and the visited" (Global Ecotourism Network, 2016).	Guided forest walks, mountain gorilla trekking, wildlife conservation volunteering holidays, Reforestation volunteering holidays
Adventure tourism	Trips that include "at least two of the following three components: a physical activity, natural environment, and cultural immersion" (Adventure Travel Trade Association, cited in UNWTO, 2014, p.10).	Walking, hiking, cycling, canoeing, kayaking, sky diving, bungee jumping, caving, rock climbing, trekking, mountaineering, snowboarding, skiing
Wildlife tourism	"Observing animals in their natural environment is the main purpose. This includes both land-based and water-based environments. The focus is on observation (wildlife watching tourism), but it can also involve interaction such as touching or feeding animals" (CBI, 2017).	Bird watching, safari tours, whale watching
Wilderness tourism	"travel to remote destinations throughout the world that may be designated wilderness, national park or other protected area" (Mintel, 2014).	Trekking, camping and caravanning, canoeing, kayaking

The debate is mainly associated with the differing tourist motivations and behaviours and the sustainable nature of activities involved. For each form of tourism listed in Table 4.1, the immersion in nature might be deep, such as in trekking and camping in a remote area for an extended period of time, or shallow, such as visiting selected natural beauty spots for a day. The main motives of visitors may vary too. For adventure tourists, participating in the physical activity may be the key motivator, rather than spending time in nature. Despite being frequently used synonymously with nature-based tourism, ecotourism is not simply travelling to a natural area for pleasure, but conservation and learning should be the main motives of ecotourists. Adventure tourism and wilderness tourism are difficult concepts to define as adventure and wilderness are subjective and socially constructed notions. Some people may also categorise consumptive activities such as fishing and hunting as wildlife tourism, whereas many would limit wildlife tourism to non-consumptive watching and interaction only (Mintel, 2019). Hence, defining the different forms of tourism is not easy.

Depending on how one defines the different types of tourism, the associated tourism products and activities will differ. Yet, in practice, nature-based holidays can include a range of activities across the different forms of tourism. For example, a holiday product labelled as 'astrotourism' can combine astronomy activities with other nature tourism activities, including outdoor adventure, wildlife watching or photography activities as well as visits to heritage sites.

An area's natural attraction or resources can differ significantly depending on natural changes in seasons and therefore it is necessary to consider the temporal dimension in nature-based tourism. Some activities rely on seasonal and weather changes (e.g., wildlife migration, winter sports) whereas others may rely on the time of the day (e.g., sunrises, northern lights), or changes in celestial cycles (e.g., eclipses). Natural tourism attractions are also affected by more long-term passing of time and changes in nature and climate. The impacts of climate change on natural resources of tourism destinations have become more visible (e.g., coral bleaching in the Great Barrier Reef). It is the focus of this chapter to integrate the concept of temporality into the nature-based tourism dialogue and how nature and time together create tourism experiences. Tourism organisations and businesses need to understand the temporal changes in natural tourism resources as they influence consumer demand and in return affect their product development and marketing strategies.

Tourism demand for nature

An increasingly digitally connected, work-centric and material world has resulted in many people seeking solace by looking inward and giving greater consideration to their own wellbeing whilst travelling (Mintel, 2019; 2021). Consequently,

> *"many travellers are turning to nature to reset and rebalance the composure that has eroded in their daily lives, as well as to find an environment that is attractive and spacious in which to physically improve their health"* (Mintel, 2019).

Consumers are also more aware of climate change and their impact on the environment, and a company's sustainability is now a key factor in their travel purchase decision, especially for young people (Mintel, 2019).

Correspondingly, nature-based tourism and the related areas of adventure, camping/eco-pods and wellness tourism are predicted to accelerate in future years (Mintel, 2019). In particular, walking and trekking that enable tourists to have close and intimate encounters with a destination's culture and natural environment are increasing in popularity (AITO, 2020). Similarly, responsible interactions with wildlife, such as swimming with sea lions in the Galapagos, kayaking through the Amazon jungle or tracking animals are becoming as popular as traditional safaris. The popularity of the traditional safari remains strong, in part due to the diversification of African safaris from simply providing travellers with photo opportunities of the big five to championing conservation and working with local villagers to protect the wildlife.

Tourism diversification also enables nature-based destinations to overcome the challenges of seasonally pronounced high and low peaks and encourage tourists to visit year-round. For example, in Africa, protected areas that offer wildlife watching tours are subjected to climatically driven seasonality (Mintel, 2019). During the rainy season, instead of focusing on wildlife watching, ecolodges can diversify their offer to wellness and adventure tourism. For example, the rainy season provides kayaking and white-water rafting opportunities, and, in some areas, rain holds deep cultural significance, and there are festivals and celebrations that tourists can experience centred on these.

Tourism, natural migrations and natural phenomena

Nature-based tourism relies on natural attractions and resources of a place, including flora and fauna and land-, water- and sky-scape. These natural resources change with the passing of time and these temporal changes in nature create tourism demand. Tourism is influenced by seasonal flora and fauna lifecycle patterns as shown in the annual wildebeest migration holidays in Africa, humpback whale migration in Australia (see Case study 1), spring cherry blossom holidays in Japan, and fall foliage (leaf-peeping) holidays in New England, USA.

Case study 1: Whale watching tourism, Australia

Australia is a popular destination for whale and dolphin watching. While dolphin watching tends to occur all year round, whale watching activities are concentrated during the whale migration seasons as the animals migrate from and to Antarctica around the country. Whale-watching tourism in Australia has gained popularity in recent years and is now an important economic opportunity to many coastal destinations (Commonwealth of Australia, 2017).

The Ningaloo Coast in the State of Western Australia is a popular location for whale watching. The area, already renowned for swimming with whale sharks, started a trial for swimming with humpback whales in 2016. The trial was expected to extend the tourism season as the humpback whale migration season almost perfectly coincides towards the end of the whale shark season (The Guardian, 2015).

Although whale watching is frequently promoted as ecotourism by destinations and operators, and regulations are in place, there are concerns about the impact of whale watching tourism on the welfare of the animals. The presence of people and repeated harassment can cause stress and behavioural changes in animals (Animal Welfare Institute, n.d.). Collisions between whale-watching vessels and whales can cause serious injury and death to the animals and in rare cases to the whale-watching participants too. Accidents have been reported where participants suffer serious injuries after being struck whilst swimming with whales. In 2020 at least three people got injured by humpback whales in two separate incidents in Ningaloo and in both instances a calf was present (Pascual Juanola, 2020).

Tourism activities are also affected by other temporally defined natural phenomena. A relevant example is astrotourism, which is:

"a special interest tourism (SIT) market, in which travellers are motivated by sky-observation related experiences and their travel/destination choices are based on these experiences" (Soleimani et al., 2019, p.2309).

It includes travelling for purposes such as, stargazing, northern light sighting, eclipse chasing and visits to observatories. Most activities are temporal by nature and dictated by time of day, month and year, and celestial cycle. Astrotourism is often considered as a form of nature-based tourism or ecotourism (Soleimani et al., 2019) as unpolluted, dark night skies are the primary resources (See Case study 2).

Case study 2: Dark sky tourism, UK

As of March 2021, ten locations in the UK have been awarded International Dark Sky Park or Reserve status by the International Dark-Sky Association (IDA, n.d.). Such designation provides the areas with an opportunity to raise awareness and develop new tourism products. These locations offer way-marked paths, guided tours and organised astronomy events, including Dark Skies Festival throughout autumn and winter (Dark Skies National Parks, n.d.). Dark sky tourism relies on unpolluted, dark night skies and can benefit rural destinations by attracting tourists, particularly outside of peak season, boosting tourist expenditure and overnight stays. Protecting the night skies can also help protect habitats for wildlife, improve health and wellbeing of the community.

In Scotland, northern latitudes and long winter nights, and expanses of unpolluted, dark skies, present tourism businesses opportunities to benefit from dark sky tourism. The Galloway Forest Park was declared Scotland's first Dark Sky Park (the fifth in the world) in 2009 and used tourism and advocacy to lead, educate and change people's attitude on light pollution (Rinaldi, 2019). In England, the Kielder Water and Northumberland National Park were awarded Dark Sky Park status by the IDA in 2013. A survey of tourism businesses in 2017 revealed that dark sky tourism created economic benefit of over £25million, supporting 450 jobs with 15% businesses reporting an improved business performance in the area (Northumberland Tourism, n.d.).

Last-chance tourism

Last-chance tourism (LCT) is a travel trend based on the temporality of vanishing tourism resources whereby tourists seek to experience them before they have irreversibly changed or disappeared entirely (Lemelin et al., 2010). Motivated by an impending sense of loss, tourists rush to visit a destination perceiving that time is running out (Fisher & Stewart, 2017). For the most part, environmental and ecological loss has driven this demand, however, an expected change in cultural resources, such as historical sites and indigenous cultures, or political changes, such as visiting communist Cuba, have also pulled tourists to a LCT destination (Lemelin et al., 2010). Furthermore, seeing a destination before it becomes too commercialised or visiting a destination while one is still fit enough to make the journey, especially as some LCT destinations are in remote areas, have also been associated with this form of tourism (Fisher & Stewart, 2017).

The concept of LCT first emerged to describe an increased tourism interest in cold regions, especially polar regions which are highly vulnerable to climate and environmental change (Dawson et al., 2011) (see Case study 3). Consequently, climate change is commonly considered in relation to LCT and the impact it is having or will have on natural tourism resources.

Case study 3: Polar bear viewing tourism, Canada

Arctic temperatures are increasing and causing longer ice-free seasons. Polar bears feed on ice-dependent seals. The longer the bears are able to feed during the winter months, the better they are able to survive during the ice-free fasting periods (Dawson et al., 2010; Lemelin et al., 2010). Churchill, Manitoba, Canada is a popular destination to see polar bears as they congregate along the shores of the Hudson Bay for approximately six weeks during the autumn, where they decrease their metabolic rates and subsist on stored fat reserves (fasting), allowing them to conserve energy while they wait for the sea ice to form (Dawson et al., 2010; Lemelin et al., 2010). Whilst the extended ice-free season provides tourists with a longer period of time to see the bears, it is also contributing to the decline in the health of the bears, particularly for female bears, as the longer fasting periods are leading to a greater reduction in weight, making them less likely to produce cubs. Aware of the impact of climate change on polar bears, tourists are motivated to view the species before it is too late.

Tourists engaging in LCT have strong pro-environmental values and they behave in sustainable ways while in the destination (Denley et al., 2020). However, the long-haul travel necessary to reach remote locations contributes to the demise of the very attraction they visit through greenhouse gas (GHG) emissions (Dawson et al., 2010; Lemelin et al., 2010). In addition, an influx of tourism can result in excessive pressure on and exploitation of an already ecologically fragile area. Therefore, the impact of LCT is partly indirect and intangible, and it is implicated by spatial and temporal lags (Dawson et al., 2011). This makes it more difficult to manage and mitigate than other forms of tourism that involve direct and local impacts.

It is unclear why there is a misalignment between LCT travellers' pro-environmental values and their unsustainable behaviour of travelling long-haul to reach LCT destinations. However, Denley et al., (2020) consider two reasons. First, whilst tourists might be well-intentioned, they are not making the connection between their LCT trips and the impact of their travel on climate change and how it contributes to the demise of the destination they are visiting. Second, tourists could be 'impact neglecters' choosing to separate themselves from the harm they cause (p. 1874). Despite this misalignment, the pro-environmental values of these tourists could be harnessed to mitigate the impacts of LCT. For example, tourists could become ambassadors for the protection of the LCT destinations and support conservation activities and organisations (Lemelin et al., 2010). Another option would be to tackle the pro-environmental attitude-behaviour gap by emphasising the amount of GHG emissions associated with LCT trips and encourage travellers to either offset the emissions from their trips or live a more environmentally friendly lifestyle back home (Denely et al., 2020).

Slowness and the value of time in nature

Appreciating the natural environment on holiday has also been linked to slow tourism. This encapsulates spatial and temporal practices, and it is underpinned by the desire to connect in particular ways and disconnect in others (Fullagar et al., 2012). Slow tourism is premised by a need to savour time, to dwell and connect with places and value travel experiences as forms of lived knowledge (Fullagar et al., 2012; Varley & Semple, 2015). It provides temporal deceleration and spatial distance from fast paced modern life which is typified by timesaving, stress, pressure and tension (Fullagar et al.,

2012; Oh et al., 2016). Key motivations include relaxation, enriching oneself, restoring and revitalising body and mind, and engaging in the environment sensuously. Slow travellers are distinguished by a desire to experience a different temporality to those who have a list of things to see while away, which they tick off and move on from (Fullagar et al., 2012). Therefore, slow tourists reject fast tourism characterised by standardised products with predicted outcomes that do not provide an opportunity for self-enrichment.

Slow immersion of place can incite different transport modalities, such as walking, canoeing and leisurely cycling (see Figure 4.1) that value nature by enabling the tourist to immerse themselves in the natural environment and use low carbon forms of transportation. Therefore, enjoying the journey is an integral part of the holiday. Connecting with communities, such as enjoying cultural traditions and local hospitality, also characterise this form of tourism. Combined, these enhance wellbeing, enable self-growth, offer a genuine connection with the destination and create a sense of meaning (Fullagar et al., 2012; Oh et al., 2016). Accordingly, slow travel practices are informed by ethical sensibilities that engage people with nature and or culture and foster a critical awareness of the impact of one's own tourist behaviour (Fullagar, et al., 2012).

Figure 4.1: Slow tourism: Cycling the Coast to Coast, England

Adventure tourism, once defined by the fast and de-natured intense experiences of risk and thrill, such as a half-day activity or a multi-activity week, now captures the simple rich experience of extended time in the natural environment (Varley & Semple, 2015). Recognising the growing trend to seek unusual new luxuries in the form of time in nature, cooking your own wild food on a wood fire and carrying your own luggage over rough lands (See Figure 4.2) or along remote coastlines in kayaks, Varley and Semple (2015) propose the concept of slow adventure tourism. In doing so they consider the temporal, natural and corporeal dimensions of being and journeying in the natural environment.

Figure 4.2: Slow adventure tourism: Backpacking in Scotland

For example, they discuss how time is felt via natural change such as the dropping of the sun, the air becoming cooler and shadows lengthening, as well as tides and weather. Being exposed to nature, especially when over an extended period of time, allows tourists to connect with nature, such as appreciating the stars in the sky without light pollution, watching a storm, sleeping outside and encounters with wildlife. Accordingly, a slow adventure journey through the natural environment can present challenges and it requires commitment from participants of both time and energy. Therefore, *"in slow adventure, time does not merely pass, but is felt, in bodily rhythms of tiredness, sleep, wakefulness and effort"* (Varley & Semple, 2015, p. 82). Although, slow adventure tourism offers temporal deceleration and a therapeutic space from hypermodernity, allowing one to dwell and connect with place, it is also through these challenges and embodied experiences that tourists learn about themselves.

Summary

The natural environment is an essential resource for many forms of tourism. Because of this, delineating nature-based tourism from other types of tourism is problematic, as their respective tourists share similar motivations and behaviours and use the same natural resources. Temporal constructs offer a lens to analyse the complex inter-relationship between nature and different forms of tourism.

Temporal changes in natural tourism resources, such as seasonal fauna and flora lifecycle patterns, natural phenomena and vanishing natural tourism resources are key pull factors to a destination for many forms of tourism. The natural environment also provides an opportunity for temporal deceleration and spatial distance from fast paced modern life. Immersion in nature enables tourists to dwell, savour time and recharge, and in return, improves wellbeing, be it physical or mental, which is a key motivator for all tourists. By immersing oneself in nature, tourists may experience an increased sense of responsibility towards the environment and become ambassadors for the protection of the destinations they have visited and the flora and fauna they have observed.

Self-reflection questions

1. Consider how nature is a key component of different types of tourism.
2. Taking a nature-based tourism destination of your choice, explain its natural resources and discuss the different forms of tourism that utilise these natural resources over the course of a year.
3. Summarise five key factors that are driving demand for nature-based tourism.
4. Animal Welfare Institute advised that whale watching operators should respect the animals first and the clients second, and tourists should not trust those operators guaranteeing a whale sighting. How can operators respect the animals and satisfy their clients at the same time?
5. Assess the positive and negative impact that last chance tourism is having on a destination of your choice.
6. Consider the role of nature in slow tourism and the benefits of slow immersion in nature for participants.

References

Animal Welfare Institute (n.d.) Whale Watching. Retrieved from: https://awionline.org/content/whale-watching

Association of Independent Tour Operators (AITO) (2020) AITO Travel Insights Report 2020. www.aito.com/media-area/press-office/travel-insights-2020

CBI (2017) Wildlife tourism from Europe. www.cbi.eu/market-information/tourism/wildlife-tourism/europe

Commonwealth of Australia (2017) Australian National Guidelines for Whale and Dolphin Watching 2017. www.environment.gov.au/marine/publications/australian-national-guidelines-whale-and-dolphin-watching-2017

Dark Skies National Parks (n.d.) About national parks dark skies. Retrieved from: https://www.darkskiesnationalparks.org.uk/about

Dawson, J., Stewart, E., Lemelin, H. & Scott, D. (2010) The carbon cost of polar bear viewing tourism in Churchill, Canada, *Journal of Sustainable Tourism*, **18**(3), 319-336.

Dawson, J., Johnston, M., Stewart, E., Lemieux, C., Lemelin, R., Maher, P. & Grimwood, S. (2011) Ethical considerations of last chance tourism, *Journal of Ecotourism*, **10**(3), 250-265.

Denley, T., Woosnam, K., Ribeiro, M., Boley, B., Hehir, C. & Abrams, J. (2020) Individuals' intentions to engage in last chance tourism: applying the value-belief-norm model, *Journal of Sustainable Tourism*, **28**(11), 1860-1881.

Fisher, D. & Stewart, E. (2017) Tourism, time and the last chance, *Tourism Analysis*, **22**, 511-521.

Fullagar, S., Wilson, E. & Markwell, K. (2012) Starting slow: Thinking through slow mobilities and experiences, in S. Fullagar, K. Markwell and E. Wilson (eds.), *Slow Tourism: Experiences and mobilities*. Bristol: Channel View, pp.15-26.

Global Ecotourism Network (2016) Definition and key concepts. Retrieved from: www.globalecotourismnetwork.org/definition-and-key-concepts/

International Dark-Sky Association (n.d.) International Dark Sky Places. Retrieved from: www.darksky.org/our-work/conservation/idsp/

Lemelin, H., Dawson, J., Stewart, E., Maher, P. & Lueck, M. (2010) Last-chance tourism: the boom, doom, and gloom of visiting vanishing destinations, *Current Issues in Tourism*, **13**(5), 477-493.

Oh, H., Assaf, A. G. & Baloglu, S. (2016) Motivations and goals of slow tourism, *Journal of Travel Research*, **55**(2), 205-219.

Mintel (2014) Wilderness Tourism – November 2014. Retrieved from: https://reports-mintel-com.hallam.idm.oclc.org/display/680926/

Mintel (2019) Wildlife refuge tourism & market differentiation – international – May 2019. https://reports.mintel.com/display/924318/

Mintel (2021) The ethical traveller – UK – February 2021. https://reports.mintel.com/display/1042549/

Northumberland Tourism (n.d.) Northumberland Dark Sky Research. https://www.northumberlandtourism.org.uk/research-insights/regional-national/dark-sky-research

Pascual Juanola, M. (2020) 'No freak accident': Scientists flagged concerns with Ningaloo humpback swimming tours as early as 2015. https://www.watoday.com.au/national/western-australia/no-freak-accident-scientists-flagged-concerns-with-ningaloo-humpback-swimming-tours-as-early-as-2015-20200819-p55n6w.html

Rinaldi, G. (2019) A decade of the UK's first Dark Sky Park in Galloway. Retrieved from: www.bbc.co.uk/news/uk-scotland-south-scotland-50405389

Soleimani, S., Bruwer, J., Groww, M.J. & Lee, R. (2019) Astro-tourism conceptualisation as special-interest tourism (SIT) field: a phenomenological approach, *Current Issues in Tourism*, **22**(18), 2299-2314

The Guardian (2015) Swimming with humpback whales to be trialled at WA's Ningaloo marine park. Retrieved from: https://www.theguardian.com/environment/2015/nov/01/swimming-with-humpback-whales-to-be-trialled-at-was-ningaloo-marine-park

UNWTO (2014) Global Report on Adventure Tourism. Retrieved from: https://www.e-unwto.org/doi/epdf/10.18111/9789284416622

Varley, P. & Semple, T. (2015) Nordic slow adventure: explorations in time and nature, *Scandinavian Journal of Hospitality and Tourism*, 15(1-2), 73-90.

5: A Chronological Exploration of Key Influences on the Development of Tourism

Neus Crous-Costa, Dolors Vidal-Casellas and Nuria Morere-Molinero

Learning outcomes

Having read this chapter, you will be able to:

1. Understand how the development of tourism has been informed by historical events.
2. Appreciate key chronological developments in providing a temporal context to contemporary tourism.
3. Understand how historical events gave rise to cultural, wellness and escapist forms of tourism.
4. Appreciate how authoritarianism regimes influence tourism behaviour.
5. Consider the concepts of Fordism and post-Fordism in shaping contemporary tourism.

Introduction

Contemporary tourism is often accepted as the direct heir of the European Grand Tour in the 18th century, which in turn succeeded the Age of Discovery during the 15th - 17th centuries (Chaney, 1998). However, a closer look into the subject reveals the existing tourism ecosystem has deeper roots (Morère Molinero, 2021). It encompassed various reasons for travel – for trade, pilgrimages and diplomacy, among other purposes. The values that emerged from the Enlightenment and the Industrial Revolutions, in addition to the gradual crystallization of capitalism, were instrumental in shaping new geopolitical structures worldwide, in which Europe occupied the hegemonic centre (Towner, 1985; Bertrand, 2008). At the same time, these major events transformed the forms and rhythms of both life and work in Europe.

The aim of the chapter is to present the sociological dimension of travelling related to the different historical events over the past few centuries to the present day. This text is therefore structured according to some of the main socio-historical events that have affected the development of tourism, following as much as possible a chronological order. Geographically, Europe was chosen as the focus of this chapter, to illustrate its role in the origin and expansion of tourism to other areas of the world. However, it is acknowledged that tourism development in other parts of the world, while influenced by Europe's growing leisure phenomenon, were also shaped by local and nearer-to-home forces.

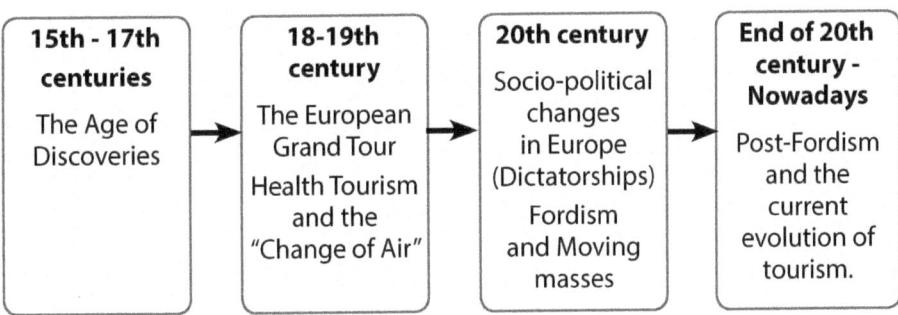

Figure 5.1: Time chronology of the events covered in this chapter

The opening up of the European Grand Tour : popularization of tourism

The Grand Tour of the European aristocracy was the initial journey of discovery into the historical and artistic treasures of Europe. The bourgeoisie and the middle classes joined this trend through Europe in the 19th century for cultural education. For example, Vidal-Casellas (2006) demonstrates the existence of what we now call cultural tourism at the beginning of the 20th century in Barcelona, which was promoted by the Society for the Attraction of Foreigners, which was a well-known organization in Europe. Likewise, such trips were considered an activity for socialization and a way to consolidate a social position.

The main destinations of the Grand Tour were located in France, Italy, and the Alps (Towner, 1985; Hornsby, 2000; Walton, 2002), while a very few cultural tourists, like Lord Byron, reached areas perceived as remote or peripheral at that time: Greece and Portugal, for example (García Mercadal, 1952; Lleo Cañal, 1984). Although it was not until the 19th century that the first travel guides appeared as we know them today, people already carried books that provided guidance on when and how to visit a destination, for example acknowledging uncomfortable local climatic characteristics.

The middle classes had fewer resources than the aristocracy for leisure trips, both in terms of time and money, and their participation in traveling to remote places was thanks to the emergence of a tourism 'industry'. Without a doubt, the work of the Thomas Cook family is the most recognized antecedent of modern-day tourism. Their first rail excursion (1841) took 500 people from Leicester to Loughborough (Armstrong & Williams, 2005). It is important to note that Cook started this business as a religious and social service, which was closely related to the alcohol abstinence 'Temperance' movement during the mid-Victorian times. Thomas Cook's son contributed with commercial knowledge and the organised 'package' journeys outside Europe followed, including to the United States (1866), Egypt and Palestine (1869) and even an around the world itinerary (1872-1873) (Swinglehurst, 1982). The company ran successively until 2007, when a family successor formed the final Thomas Cook business which went bankrupt in 2019 after 178 years of history. The brand was subsequently purchased by a Chinese group and has since been re-established as an online operator.

Yet, Cook's family and company were not free from criticism. For example, Walton (2010) mentions that Cook has gained disproportionate prominence within tourism studies, and that the founders were largely unknown. Innovations such as hotel and restaurant coupons were already being used in the 1840s by Henry Gaze, a contemporary and key rival of Cook. Walton states that Cook was not an innovator in tourism, rather a guide among the popular classes, since most of his clients belonged to social classes that otherwise could not have travelled (for example, unmarried women, lower ranks of the clergy…) or appreciated what they saw without guidance. The London press was quick to react to Cook's Tours, often snobbishly, pointing to a loss of 'exclusivity' within the leisure travelling experience.

Figure 5.2: Thomas Cook travel programme, Thomas Cook Archive: Record Office for Leicestershire, Leicester and Rutland

Simultaneously, the intangible heritage of non-European civilizations started to be attractive to tourists and became fundamental for the development of anthropology and psychology, as well as literature. A few examples are Hermann Hesse the German-Swiss poet and novelist whose works commonly explored self-knowledge and the quest for spirituality; Carl Gustav Jung, and the French anthropologist and ethnologist Claude Lévi-Strauss who created the famous phrase:

> *Je hais les voyages et les explorateurs et voici que je m'apprête à raconteur mes expedition'*. ('I hate travelling and explorers. Yet here I am proposing to tell the story of my expeditions'). (Lévy Strauss, 1955, p. 10).

> **Case study : The continuity of expeditions**
>
> Within a Western lens, 'Modernity' is claimed to have begun with the Age of Discoveries (15th - 17th centuries). Many travellers accumulated natural specimens and cultural objects from their places of origin, including among others Ruy González de Clavijo (traveller, writer and Castilian Ambassador to the central Asian Timurid Empire) and Juan Sebastián Elcano, the first known circumnavigator of the world with Magellan in 1519-22. In the following centuries, expeditions involved systematic plundering of host societies' cultural artifacts, which in turn furthered the development of science (especially natural science) and culture in the European and North American metropolises. One of the most famous examples was the late 19th and early 20th century archaeological campaigns of Pharaonic Egypt by Howard Carter, which were underwritten by the Earl of Carnavon, a British aristocrat collector of antiquities Although this type of journey is not usually considered as a 'tourism' activity in the collective awareness, these trips were technically touristic in nature, since they depended on the use of infrastructures such as local transport, accommodation and guiding services. Moreover, renewed interest in 'Egyptology' became a distinct form of tourism in the ensuing decades (Starkey & Starkey, 2001; Moussa, 2004; Anderson, 2012).

Nowadays there are regulations that prohibit the export of valuable objects from the visited country, thus looting is often linked to wars. Afghanistan and Iraq are two recent examples, as well as the areas occupied by the Islamic State in the early 21st Century (Pappa, 2018). However, there is a deep public debate about the legitimacy with which Westerners travel to other countries to learn from their lifestyle, which they then capitalize in their places of origin. Recently, some studies have been carried out about the negative effects of cultural appropriation for film creation and its effects (Chen et al., 2019).

Health tourism for the mind, body and spirit

Two negative externalities of the Industrial Revolution were the living conditions of the working classes and the increasing pollution of cities, both air and water. This, combined with the values of the Enlightenment (including scientific advances) changed the perception of nature from sacred and dangerous to utilitarian. This process of changing perceptions also had its impacts on tourism, giving rise to new forms of recreational travel.

Thermalism, baths and 'hygiénisme'

Heeley (1981) describes how spas started to be built in inland areas during the 16th century and eventually expanded to include resorts in coastal areas as places for both health treatments and socialization. They were developed intensively in the 19th dedicated to wellness purposes along many European coasts. Some of them such as Biarritz in France or San Sebastian in Spain were frequented by royalty, which benefited the entire municipality in terms of improvements in construction and infrastructure (Laborde, 2002; Larrinaga Rodríguez y& Pastoriza, 2009; Cirer-Costa, 2014). Other centres such as in Sant Feliu de Guíxols (Spain), had two separate treatment areas for different customers, to segregate locals and visitors.

In times of antiquity and until the 17th century, the ocean was often considered a dangerous place, despite its utility for fishing and commerce. The coastal landscape was a liminal space that aroused trepidation, since it was interpreted as a place where cataclysm and order found balance (Corbin, 1994; Romm, 1992). Over time, such maritime mythologies gradually disappeared thanks to the discoveries of maritime routes, the expansion of trade, scientific advances and the development of Enlightenment culture. By the 19th century, the oceans were being harnessed for maritime tourism.

A parallel trend gathered pace with the development in the mid-19th century of 'hygiénisme', related to breakthroughs in medical science and microbiology (Martín de la Torre, 1996; Lindemann, 1999). The work of Louis Pasteur in advancing understanding of the causes and prevention of disease was particularly significant and ultimately impacted on aspects of urban design, leading to improvements in the health of urban areas and their inhabitants. Accordingly, urban areas started to develop neighborhoods focused on preserving physical and mental health through recreation, including gardens and orchards, sewage systems and running water, among other innovations. This awareness also shaped the way in which food and many other aspects of life were perceived. Travel and vacations were increasingly related to health. For instance, sea and mountain air were frequently prescribed as treatments for patients with diseases such as tuberculosis in Europe (Guillaume, 1991). 'Taking the waters' became associated with defined travel-seasons, reflective of climatic patterns in the emerging health destinations (Gil de Arriba, 1996; Díaz Cano & Pérez Redondo, 2015).

Mountain hikers

The establishment of the Alpine Club in London in 1857 paved the way for a continuous increase in the number of hiking societies in Europe. The rediscovery of the rural environment soon started to integrate scientific and humanistic activities, since many of its practitioners showed interest in topics such as natural sciences, archeology, art history, ethnology and/or anthropology (Martí-Henneberg, 1986). Roma i Casanovas (2011) explains that the discovery of mountains for recreational purposes was not a linear historical process and that its sociocultural context is often ignored. Like the changing perceptions of maritime environments, the culture and increasing secularization in some societies changed the social value of the mountain. It was no longer considered a distant place of hierophany (i.e. divine, sacred or mythologised). Rather, mountains became increasingly seen as healthy places in which emotions for the landscape as spectacle, adventure and knowledge were shaped. This new approach slowly contributed to reversing the utilitarian view of nature, to one of a place of encounter with oneself and a place of inspiration for artists and writers (for example, the 19th century American naturalist Henry David Thoreau and the German Romantic landscape painter Kaspar David Friedrich).

Generally, individuals initiated this activity by visiting the periphery of their cities of residence and then gradually widening the distance until reaching mountain ranges further afield, especially in other countries. The Alps became one of the most coveted destinations to the point that John Ruskin (a 19th century English artist and social thinker), who wrote a lot about the beauties of the Alps and who dedicated great efforts in promoting hiking and tourism among the working classes, lamented deeply about the damage visitors caused there. He strongly criticized the developments aimed at providing comfort to tourists. A casino was built in Chamonix and for that, he lamented English society for teaching the Swiss *"the foulness of the modern lust of wealth ..."* (Ruskin cited by Hanley & Walton, 2010, p.99). By the 1920s, the summertime recreational activities and respite were supplemented with 'winter sports tours', actively promoted by the Thos. Cook company, Sir Henry Lunn (founder of Lunn Poly tours) and Alpine clubs in various European countries (Swinglehurst, 1982).

Authoritarian regimes, international visitors, and national tourism

The 20th century witnessed major events that reconfigured European borders and caused major socio-political changes. After various revolutions and war conflicts, some countries fell under totalitarian governments. Some of them, despite maintaining an autarchic (autocratic) regime that restricted the opportunities for their citizens to travel abroad, acknowledged the benefits of receiving foreign tourists.

There were two main objectives for regimes like the Soviet Union (1922-1991) and the Franco regime in Spain (1940-1975) in allowing inbound tourism. First, the inflow of foreign exchange boosted their economies and second, the impressions that visitors would take back to their places of origin, helped to create a positive image of the political regime. Accordingly, it was essential to ensure that tourists had a pleasant experience. Here is where the tourists' activity met cultural diplomacy.

Yet these regimes tended not to contemplate that the arrival of foreign visitors would encourage interactions with the local population and the exchange of values (Cohen, 1984; Shaw & Williams, 2002; Pearce, 2005). Classic examples are the introduction of new ways of dressing (such as bikinis or miniskirts) or ways of socializing. While residents were used to social conservatism preserved by their governments, visitors from democratic countries were generally more open-minded (Fuentes Vega, 2017). This applied to less visual aspects as well, such as ways of thinking. For example, in the case of Spain, the years following the 1960s tourist boom are known as *el destape* (the uncover) era, alluding directly to the way women started to adapt their mode of dress and indirectly, to changes in the way of thinking. Foreign tourism thus heralded a gradual demonstration effect on Spanish society.

Possibly, as Vidal-Casellas states (in Gibert, 2012), if the regimes had known how to anticipate these effects on residents, the entry of foreigners would not have been allowed, at least at that time. North Korea learned this lesson. While international tourism is cautiously welcomed albeit strictly controlled there, interactions between foreigners and residents are not allowed. Opportunities for encounters between foreign visitors and locals (such as walking around the street, taking the subway, visiting an amusement park) remain highly restricted and guarded.

With regards to domestic tourism, various objectives were set by the regimes. The most common goals were knowledge and reinforcement of a sense of national identity, leading to the cohesion and uniformity of citizens. In the Soviet Union, Assipova and Minnaert (2014) show how tourism was also associated with health as well as with the improvement of the individual for the sake of the political ideals of the State.

Moving masses : Fordism and Post-Fordism

Fordism

The Fordist period is one of the most studied in the history of tourism, due to its temporal proximity and the societal changes that it encouraged. Fordism signifies an intensification of production, productivity and attendant social and political structures, characteristic of the era of advanced capitalism (Watson, 2019). The period characterized by Fordism in tourism extended from the end of World War II (1945) to the late 1980s and arguably beyond. The intensification of tourism was very important in Spain (Pack, 2006), particularly with the example of resorts such as Benidorm (Ivars i Baidal et al., 2013; Nolasco-Cirugeda et al., 2020). From this time onwards, the middle and working classes in Europe began to travel 'en masse' often with all-inclusive 'packages', for up to several weeks at a time for rest and recreation, typically to the rapidly emerging 'sun, sea and sand' destinations. One of the main destination areas was the Mediterranean, due to its climate and easy accessibility. Various means of transport (coach tours, continental rail travel and especially charter flights) enabled northern European countries to expand this type of tourism establishing seasonally-defined north-south leisure travel migrations (Bramwell, 2004).

Packaged trips and product standardization were the hallmarks of this period for Europeans. The 'escape behaviour' was not surprising for populations that had endured two wars that affected a substantial part of the globe, with subsequent postwar events. In 1948 the recently formed United Nations Organisation published the Universal Declaration of Human Rights. Its article 24 specifically recognizes the right to rest and the right to have paid non-workdays to enjoy free time. Much of Europe recognized these rights but still in 2022, they are not legally guaranteed in some nations including the United States, where paid holiday entitlement depends on each enterprise (Li et al., 2022).

Post-Fordism

Although the model of Fordist mass tourism was predominant in the post-war years, it coexisted with various forms of independent travel. One of the best-known examples was the Hippie Trail, particularly evident between the late 1950s and late 1970s. It was known by many young Europeans who traveled overland to India or Nepal, in search of spirituality, adventure and alternative lifestyles (Sobocinska, 2014). These travelers also joined the Asian diaspora in the dissemination of Eastern philosophies and practices in the West, including yoga, meditation and martial arts, among others. During those decades, young people who also made shorter trips to the Mediterranean and/or extended travel to Southeast Asia, were later denominated as 'backpackers' (Sorensen, 2003; Martín-Cabello, et al., 2017).

Since the 1980s, various social and cultural changes, specially in demand motivations, took place that transformed Fordist demand into more individualized patterns (MacCannell, 1999). Consequently, tourism demand increasingly splintered into various segments that have remained prevalent to this day, for example, alternative tourism as cultural, nature-activity based, ecotourism, museums, creative tourism (Richards, 1996; Wearing & Neal, 1999; Wearing, 2009; Richards, 2011), or urban tourism such as city breaks, festivals, contemporary art (Origet de Cluzeau, 2013; Kadri & Khomsi, 2017). They are all focused on the experience. However, many touristic trips are multipurpose in practice. In other words, people carry out activities classified in more than one niche during the same vacation (Perelló & Morere, 2020).

The new millennium fomented the growth of low-cost airlines. Over the years, the significant expansion of international travel they have facilitated has impacted on the destinations served, both socially (for example, anti-social tourist behaviours) and economically (for example, low per capita local spending). Debates on the granting of public subsidies, on the impacts of aviation on the environment and on passenger rights have resulted. It is difficult to separate low-cost aviation from the accelerated gentrification processes in many cities and from claims of environmental 'greenwashing'. At the same time, this situation coexists alongside the significant reduction in flight prices which has allowed society to discover first-hand a Europe that has long valued social cohesion. 'Experience' has become a potent construct in many peoples' travel behaviours. Similarly, discovering or travelling 'off the beaten path' has become increasingly valued, leaving behind/escaping from the 'mass'.

Before the global financial crisis in 2008, many Europeans had joined the trend in travelling as often as possible, for example on short trips to European cities, on weekends and/or longer trips to other continents, typically in summer, Easter and Christmas 'vacation periods'. Tourism was no longer just a human need or an individual desire; over time it has become an inescapable social requirement.

In this sense, it must be acknowledged that more and more critical voices are being raised, not only against the urban-social and ecological impacts of tourism, but also against its motivations. For example, in the 21st century, commercially driven volunteer tourism and solidarity tourism has both flourished and become the target of more objections, with the argument that it is focused more on satisfying the altruistic needs of the 'First World' than on actually alleviating global inequalities. (Fuentes-Moraleda et al., 2016)

Summary

This chapter has approached temporality in tourism from a historical perspective, chronologically reviewing aspects of the development of tourism since the Grand Tours of the 18th century and examining how travel and living conditions influence each other, while considering the social value of travelling.

A common theme is that travelling encourages personal cultural exploration. This idea was already present in the Grand Tour, and it continues nowadays with various forms of tourism. At the same time, it is inevitably associated with post- or neo-colonial structures as well as with debates such as the appropriation of intangible heritage and of international tourism by authoritarian regimes.

Importance has also been given to the relationship with nature. The unsanitary conditions produced by the Industrial Revolution heralded a new approach to nature for pleasure and health purposes, which encouraged the creation of a wellness 'niche' in the tourism industry, and which subsequently has raised questions about conservation and real estate developments, especially on the coasts and in mountain landscapes.

Finally, the vacation context went from reflecting societal escape from war exhaustion in the mid-20th century giving rise to industrial-scale or Fordist tourism, to the post-Fordist desire for self-discovery and exploring the world, which carries many ethical and moral considerations.

This analysis has not set out to be totally comprehensive. Priority has been given to major social events linked to tourism and some of them have only been touched upon. Culturally, behaviourally and structurally, tourism continues to evolve. In chronological terms, the period starting 2020, with the global pandemic shutdown and re-emergence, will necessitate new reflections on the social and cultural values of tourism.

Self-reflection questions for students

1. Find examples of how tourism is related to the misappropriation of tangible goods and intangible knowledge of other cultures.
2. How has the change in the cultural perception over time of the oceans/maritime environment been related to the development of sun and sea tourism?
3. What consequences does international tourism carry in places with authoritarian regimes?
4. Choose one of the topic threads in this chapter (for example Fordism, cultural tourism, health tourism, exploration). Considering recent history, how would you describe its most recent evolution?

References

Anderson, M. (2012) The development of the British Tourism in Egypt 1815-1850, *Journal of Tourism History*, 4 (3), 259-279

Armstrong, J. & Williams, D.M. (2005) The steamboat and popular tourism, *The Journal of Transport History* 26(1), 61-77.

Assipova, Z. & Minnaert, L. (2014) Tourists of the world, unite! The interpretation and facilitation of tourism towards the end of the Soviet Union (1962-1990), *Journal of Policy Research in Tourism, Leisure and Events*, 6(3), 215-230.

Bertrand, G. (2008) *Le Grand Tour revisité. Pour une archéologie du tourisme. Le voyage des Français, milieu XVIIIe siècle-début XIX siècle*, CEFR, 398.

Bramwell, B. (2004) Mass tourism, diversification and sustainability in Southern Europe's coastal regions, in Bramwell, B. (ed), *Coastal Mass Tourism. Diversification and Sustainable Development in Southern Europe*, Bristol: Channel View Publications.

Chaney, E. (1998) *The Evolution of the Grand Tour. Anglo-Italian cultural relations since the Renaissance*, London: Routledge.

Chen, R., Chen, Z. & Yang, Y. (2021) The creation and operation strategy of Disney's Mulan: Cultural appropriation and cultural discount, *Sustainability*, 13, 2751: https://doi.org/10.3390/su13052751

Cirer-Costa, J. C. (2014) Spain's new coastal destinations. 1883–1936: The mainstay of the development of tourism before the Second World War, *Annals of tourism research*, 45,18-29.

Cohen, E. (1984) The Sociology of tourism: Approaches, issues, and findings, *Annual Review of Sociology*, 10, 373-392.

Corbin, A. (1994) *The Lure of the Sea. The discovery of the seas in the Western World*, Berkeley: University of California Press.

Díaz Cano, E. & Pérez Redondo, R.J. (2015) *Sociology of Tourism and Leisure Teaching materials*, Madrid, Ommpress.

Fuentes-Moraleda, L., Muñoz-Mazón. A. & Rodríguez-Izquierdo, S. (2016) El turismo solidario como instrumento de desarrollo: un estudio de caso para analizar las principales motivaciones de los turistas solidarios, *Cuadernos de Turismo*, 37, 227-242, DOI: http://dx.doi.org/10.6018/turismo.37.256221.

Fuentes Vega, A. (2017) *Cultura visual del "boom" en España*, Madrid: Cátedra.

García Mercadal, J. (1952) *Viajes de extranjeros por España y Portugal*. Madrid.

Gibert, A. (2012) *Pensió Completa* [documentary film]. Lloret de Mar: Regidoria de la Dona, Ajuntament de Lloret de Mar.

Gil de Arriba, C. (1996) Les vacances du corps. Établissements balnéaires et activités de loisir sur la côte nord de l'Espagne de 1868 à 1936, *Annales de Géographie*, 589, 257-278.

Guillaume, P. (1991) Tuberculose et montagne. Naissance d'un mythe, Vingtième Siècle. *Revue d'histoire*, 30, 32-39.

Hanley, K. & Walton, J.K. (2010) *Constructing Cultural Tourism: John Ruskin and the tourist gaze*. Bristol : Channel View Publications.

Heeley, J. (1981) Planning for tourism in Britain. An historical perspective. *Town Planning Review* 52(1), 61-79.

Hornsby, C. (2000) *The Impact of Italy. The Grand Tour and beyond*, London: British School at Rome.

Ivars i Baidal, J. A., Rodríguez Sánchez, I. & Vera Rebollo, J. F. (2013) The evolution of mass tourism destinations: New approaches beyond deterministic models in Benidorm (Spain), *Tourism Management*, 34, 184-195.

Kadri, B. & Khomsi, R. M. (2017) The cultural and tourist city: the new face of globalization?. methaodos.*Social Science Journal*, 5(1), dx.doi.org/10.17502/m.rcs.v5i1.154

Laborde, J.-P. (2002) Nacimiento y desarrollo del turismo en Biarritz durante el Segundo Imperio, *Historia contemporánea*, 25, 51-64.

Larrinaga Rodríguez, C. & Pastoriza, E. (2009) 'Dos balnearios atlánticos entre el fin de siglo y la crisis del treinta, Sebastián y Mar de Plata: un estudio comparativo', *Historia contemporánea*, 38, 277-319.

Lévi-Strauss, Cl. (1955) *Tristes tropiques*, Paris: Plon

Li, Q., Knoester, C. & Petts, R.J. (2022), Attitudes about paid parental leave in the United States, *Sociological Focus*, 55(1), 48-67.

Lindemann, M. (1999) *Medecine and Society in Early Modern Europe*, Cambridge: Cambridge University Press.

Lleo Cañal, V. (1984) España y los viajes románticos, *Estudios turísticos*, 83, 45-53.

MacCannell, D. (1999) *The Tourist. A new theory of the leisure class*, University of California Press.

Martí-Henneberg, J. (1986). La pasión por la montaña. Literatura, pedagogía y ciencia en el excursionismo del siglo XIX, *GeoCrítica*, 66, 7-38.

Martín de la Torre, I. (1996) 'igienismo y sociedad en la España del siglo XIX, *Aportes: Revista de Historia contemporánea*, 11, 30, 3-12

Martín-Cabello, A., Anta Félez, J.L., García-Manso, A. & Pérez Redondo, R.J. (2017) *Turismo mochillero. Una aproximación desde la sociología y la antropología: una subcultura global*, Oviedo: Septem.

Morère Molinero, N. E. (2021) *Viajes culturales en la Antigüedad y el advenimiento del turismo*, Madrid: Ramón Areces.

Moussa, S. (2004) *Le voyage en Egypte. Anthologie de voyages européens de Bonaparte à l´occupation anglaise*. Paris: Laffont.

Nolasco-Cirugeda, A., Martí, P. & Ponce, G. (2020) Keeping mass tourism destinations sustainable via urban design: The case of Benidorm, *Sustainable Development*, 28 (5), 1289-1304.

Origet du Cluzeau, C. (2013) *Le tourisme culturel. Dynamique et prospective d'une passion durable*, Louvain la Neuve : Ed De Boeck.

Pack, S.D. 2006 *Tourism and Dictatorship. Europe´s peaceful Invasion of Franco´s Spain*, Palgrave Macmillan.

Pappa, E. (2018) Depoliticizing archaeology for constructing pasts and presents: Cultural heritage, war, and the West, *Radical History Review*, 130, 9-43.

Pearce, P.L. (2005) *Tourist Behaviour. Themes and Conceptual Schemes*, Bristol: Channel View Publications.

Perelló, S. & Morère, N. (2020) Museums and tourism in World Heritage Sites in Spain, in F. Cravidao and N. Santos (eds) *Management of World Heritage Sites, Cultural Landscapes and Sustainability*, Cambridge Scholars Publishing.

Richards, G. (1996) *Cultural Tourism in Europe*, Wallingford U.K.: CABI.

Richards, G. (2011) Creativity and tourism: The state of the art, *Annals of Tourism Research*, 38 (4), 1225-1253.

Roma i Casanovas, F. (2011) Del mito al monte: la conquista cultural de los Pirineos desde la vertiente sur, in *À la découverte des Pyrénées/El descubrimiento de los Pirineos*. Ville de Lourdes & Ayuntamiento de Graus. Huesca: Gráficas Alós, 43-102.

Romm, J. (1992) *The Edges of the Earth in Ancient Thought. Geography, exploration and fiction*, Princeton-New Jersey.

Shaw, G. & Williams, A. M. (2002) *Critical Issues in Tourism. A geographical perspective*, 2nd ed. Blackwell.

Sobocinska, A. (2014) Following the 'Hippie Sahibs,: Colonial cultures of travel and the Hippie Trail, *Journal of Colonialism and Colonial History*, 15 (2).

Sorensen, A. (2003) Backpacker ethnography, *Annals of Tourism Research*, 30 (4), 847-867.

Starkey P. & Starkey, J. (2001) *Travellers in Egypt*, London: Tauris.

Swinglehurst, E. (1982) *Cook's Tours: the Story of Popular Travel*, Poole: Blandford Press

Towner, J. (1985). The Grand Tour. A key phase in the history of tourism, *Annals of Tourism Research*, 12, 297-333.

Vidal-Casellas, D. (2006). *L'imaginari monumental i turístic del turisme cultural. El cas de la revista Barcelona Atracción*. PhD thesis. Girona: University of Girona.

Walton, J. K. (2002) Aproximación a la historia del turismo en el Reino Unido, siglos XVIII-XX, *Historia Contemporánea*, 25, 65-82.

Walton, J. K. (2010) Thomas Cook: Image and reality. In Butler, R.W. & Russell, R. (ed.), *Giants of Tourism*. Wallingford: CABI, 81-91.

Watson, D. (2019) Fordism: a review essay, *Labor History*, 60 (2), 144-160.

Wearing, S. (2001) *Volunteer Tourism: Experiences that make a difference*, Wallingford: CABI.

Wearing, S. & Neil, J. (1999) *Ecotourism: Impacts, potentials, and possibilities* (2nd ed.), Oxford: Butterworth-Heinemann.

6 Night and Light: Nocturnal Tourism

Raquel Cambrubí, Lluis Coromina and Jaume Guía

Learning outcomes

After reading this chapter, you will be able to:

1. Understand diurnal and nocturnal dimensions of touristic and recreational activity.
2. Understand the importance of light phenomena in urban and natural spaces.
3. Appreciate the policy implications of night-time recreational activity and artificial lighting.
4. Gain knowledge of tourism products and experiences based on attributes of darkness.

Introduction

In this chapter we deal with daily time cycles in relation to tourism, in particular differentiating the temporal variations between day-time and night-time. The day is the time normally considered for action and the night for sleep, or at least for rest. Therefore tourism activity is usually concentrated during day-time. Consequently, the taken-for-granted and dominant images of tourism destinations are most often represented in the form of day-time images, when the natural light of the sun bathes every corner of the visual landscapes. This leaves night-time as a secondary player, if at all represented, in tourism destinations' product, marketing and branding strategies.

This chapter focuses on the lesser known nocturnal reality of tourism and tourism destinations, underlying the opportunities that night-time offers to produce innovative tourism activity and enhance the attractiveness of destinations. However, the concept of 'night-life' can carry a negative connotation, characterised by irresponsible behaviours or undesirable activities of some visitors, interfering with the temporal rhythms of many local residents, for whom the night is the default time to rest, sleep and recharge from their day-time work. Therefore, well-thought policies are needed that regulate night-time recreation and make the needs of both tourists and residents compatible, to control activities that are not desirable and ensure the safety of people that seek to enjoy the nightscapes of our cities.

In the following sections, we first discuss urban tourism and night-time recreation. The power of light as an attraction is then explored, considering both artificial and natural light, followed by an exploration of 'nightscapes' as touristic experiences. Finally we consider some concerns about night-time tourism activity and regulatory policies. Two illustrative cases are presented to support the chapter: 'Llum Barcelona', and the 'London Night Time Commission'.

Urban tourism and night-time recreation

Urban tourism has witnessed a significant increase in scale during the last few decades. Reasons for visiting cities are varied. Edwards et al. (2008) state that there are a great variety of products and experiences for people with a wide range of motivations, preferences and cultural perspectives in the consumption of the urban space. Monuments, historic quarters, museums and art galleries, zoos and festivals are some of the many resources that can attract tourists to particular cities, with an amalgam of tourist activities that assist their discovery and enjoyment, services such as thematic guided tours, sightseeing bus tours. However, according to Ashworth and Page (2011: 5)

> "…the tourist city is not necessarily a distinct spatial entity that the visitor can easily recognise: it is a patchwork of consumption experiences, spatially-dispersed and often grouped into districts and zones (e.g. the entertainment zone) with symbols, a unique language and range of icons to differentiate the experience of place consumption. In this respect the tourist city is a series of sub-systems interconnected by the pursuit of pleasure, the consumption experience and a defining characteristic – the discretionary use of leisure time" (Ashworth & Page, 2011: 5).

Therefore, following this idea, tourists also make use of facilities and resources that are not necessarily touristic in essence, such as cinemas, theatres and concert venues, shops and shopping malls, bars, restaurants and nightclubs and so on. Consequently, tourists coexist within the same spaces as local residents.

When the sun goes down, tourist activity continues in urban spaces. Night-time represents a temporal variation that in some ways does not stop activity but influences and transforms the motivations and preferences of individuals. Academic literature often associates nightlife to drinking, eating, theatre and other forms of entertainment (Eldridge & Smith, 2019) and is frequently related to some socially undesired effects such as binge drinking, noise and anti-social behaviour (Eldridge, 2019).

Nevertheless, some traditional daytime activities have progressively started to extend into late night (Eldridge & Smith, 2019), for instance, the nocturnal opening of some tourism attractions such as museums, art galleries or monuments. Some cities have started to lengthen the duration of events and festivals to encompass night-time or to organise specific festivals during the night, such as light festivals. Meanwhile cities such as London have developed an extensive array of night-time activities such as late-night shopping, night bus and boat tours, thematic guided tours, theatre shows and live music. More innovative nocturnal activities are increasingly commonplace such as sleepovers at London's Natural History Museum with nocturnal dinosaur sound effects, or sport activities such as ice-skating or climbing, in addition to the traditional night-time activities related to eating and drinking. Together with all these activities, wandering the night-time city streets and sightseeing the main tourism monuments and illuminated attractions can be an attractive activity for tourists, who want to experience the city from a different gaze. For instance, Visit London offers several walking trails to enjoy the nocturnal landscape of the city and its main monuments. Considering the significance of lighting to the nocturnal urban experience, we now focus specifically on this aspect of night-time urban tourism.

Understanding the power of attraction of light

When it first entered commercial use, the role of electricity in public spaces was mainly focused on maintaining order and safety. However, it was gradually adapted for other purposes. It was as early as 1937 when the main Parisian monuments began to be illuminated (Bourgeois, 2002), extending this practice to the present day. For Demers (2010), the lighting of the urban environment allows the city to be reinvented and becomes an essential element of the urban landscape. In other words, the lighting of certain spaces or elements of a city becomes an intrinsic feature of these places. According to Bourgeois (2002), this provides the city with an image of universal value that is recognized by all. So, the tourist who visits New York may wish to see the lit up skyline of the city from the Brooklyn Bridge, or the tourist who travels to Paris will go to Trocadero Square to gaze at the Eiffel Tower illuminated with all its splendour. Camprubí and Coromina (2019a) accordingly concluded that lighting of attractions is a relevant feature of perceived image of physical elements of a destination. Therefore, the use of light in cities and applied to monuments and tourist attractions stimulates the creation of a more positive and solid image of the city, and especially of tourist spaces.

Mantei (2012) proposes the existence of three different types of illumination in urban spaces. First, there is permanent illumination, which is characterized by the continued lighting of a city, a quarter or a monument all year round. Second, seasonal illumination, which is characterized by its regular periodicity in particular time intervals during the year, for example, Christmas lighting. The Blackpool Illuminations in the British seaside town of Blackpool during November represents a longstanding example of this. Third, ephemeral illumination consists of the celebration of different types of festivals and spectacles that have light as their basis. Ephemeral illumination has been postulated as being innovative in its ability to increase the attractiveness of a tourist site over a short period of time (Camprubí & Coromina, 2019b). Although permanent illumination is frequently related to having to deal with darkness and extend the diurnal period, lighting is especially useful to beautify and valorise monuments and urban landscapes, as mentioned above.

In the case of seasonal and ephemeral illumination, the contrast of light and darkness is necessary to guarantee its success. According to Cheverier (2019: 443) *"nocturnal ritual activities create new landscapes and new sites, based on the balance between light and dark"*. Therefore, a light festival cannot be conceived without the continuous and concomitant complementarity of light and darkness. This can be seen particularly in traditional cultural festivals such as Saint Lucy's Day in Scandinavia or the Bala Chaturdashi Festival in Nepal. Contemporary light festivals in cities across the world are conceived as a tool to increase competitiveness and attractiveness (Camprubí &

Coromina, 2019b) and are perceived by local authorities as a valid alternative to traditional night-time economies (Giordano & Ong, 2017). Lyon's *Fête de la lumière* originated in 1852 and is considered the pioneer of light festivals as it has featured many of the technical advances or artistic creations for the first time. Today, the *Fête de la lumière* has served as an inspiration to other cities around the word to develop their own light festivals, for example Melbourne, Barcelona and London. Among the various types of spectacles, "son et lumière" spectacles have become highly valued for their power of attraction and their capacity to provide rich experiences such as mapping projections on the walls of architecturally distinctive buildings and combining the projected images with music and sounds (Case study 1).

Figure 6.1: Rouen Cathedral Nocturnal Light Show. Source P. Goulding

Case study 1: Llum Barcelona

Llum BCN (www.barcelona.cat/llumbcn/en/) is an event organised by the City Council of Barcelona. The festival has been held annually in mid-February in the city of Barcelona since 2012, on the occasion of the city's Winter Festival in honour of its co-patron Santa Eulàlia. The city's old quarter is filled with a variety of art installations over three days, during which night and light are partnered in a light show which illuminates and colours the city's main landmarks, where the show seeks interaction with the public by linking artistic expression, creativity, technology and heritage together.

The initiative brings together tradition and artistic 'new forms of expression' in which light is the protagonist in the city centre, transforming iconic buildings and courtyards of the city blocks' interiors. It involves the participation of various entities of the city such as the Professional Association of Lighting Designers, the Institute of Photonic Sciences, and teachers and students of various design schools. The involvement of teachers and students from the city's schools of art, interior design, architecture and lighting gives the event a more participatory dimension and contributes to promoting local, talented artists, students and teachers.

The original aim of the festival in Barcelona was to introduce residents and visitors to lighting as a growing art form by lighting up the city's landmarks and, at the same time, give residents the opportunity to rediscover the city.

In 2015, the city of Barcelona proposed three routes that allow visitors to discover the more than 20 facilities that are part of the festival. The festival was used as a starting point for the International Year of Light in 2015, which the city and the Institute of Photonic Sciences were holding at the suggestion of the United Nations. Since then new routes have been proposed every year.

Visitor numbers have increased steadily year after year, rising to an estimated 190,000 visitors at the 2019 event[1]. More than 40 light shows were shown, with creations signed by award-winning audio-visual artists with international prestige, outstanding local authors and montages from schools of art, design, lighting and architecture.

[1] Information can be accessed at:
www.lavanguardia.com/local/barcelona/20200212/473485636189/guia-espacios-instalaciones-festival-llum-bcn-poblenou-barcelona-2020.html

Tourism and natural nightscapes

It is also relevant to point out the attraction power of natural light phenomena. The light from the sky between full night and sunrise or between sunset and full night produced by the diffusion of sunlight through the atmosphere and its dust, provide yet another distinctive visual landscape. With its uniqueness and short life during the timeframe of a single day, dawn and dusk can produce high experiential value to visitors, embodied in the sunrise and sunset sky and landscapes in certain locations. Being able to enjoy the dawn or dusk from privileged places is an experience that has often been associated with romanticism. The tourist anticipates the enjoyment of the transformation from darkness to brightness, or, in the opposite direction, to see how darkness swallows up the day. The places are very varied and can be in both natural and urban environments. Some of the emblematic places that attract numerous tourists to enjoy sunsets and sunrises are, for example, the lighthouse at Cap de Barbaria in Eivissa (Spain); the Grand Canyon in Arizona (USA); Ammoudi Bay in Santorini (Greece); Mount Tamborine in Queensland (Australia) or the view from the top of the Empire State building in New York (USA).

Other natural phenomena that arouse intense interest of tourists are the Aurora Borealis (Northern Lights) and the Aurora Australis (Southern lights). These are *"a natural display of lights caused by charged solar particles entering the atmosphere, typically observable in the Polar Regions"* (Jóhannesson & Lund, 2017: 183), during the winter period. In recent years, Northern and Southern lights have been used to create tourism products, being a key ingredient for developing winter tourism in these destinations. According to Jóhannesson and Lund (2017: 183), in the case of the Northern Lights, the products have been *"sold as a mystical and romantic experience that contributes to a particular image of the North as mysterious and magical"*. In the same vein, astro-tourism has also started to be developed in some rural areas, creating products that combine activities such as sky observation and dinner, for example at the Montsec Observation Centre in the foothills of the Pyrenees mountains in northern Spain.

In general terms, lighting management has become an important issue for cities over the world. Mantei (2012) conceives the *"mise en lumière"* as a key strategic process to city tourism development, which can benefit both tourists and residents. But as we have seen, natural light and darkness can also have an important role as tourist 'pulls'. Hence the preservation and management of spaces that are free from the pollution of artificial light. These spaces are called Dark-sky Preserves or Dark-sky Reserves, and they are typically located in the surroundings of an observatory or a park aiming to promote activities related to astronomy and observation of natural light phenomena within a nightscape. According to Mitura et al. (2017) and Weaver (2011) these reserves have the capacity to develop a more sustainable tourism by limiting light emitting development within their vicinity. In such locations, lighting management is crucial for maintaining place attractiveness, offering unique visitor experiences.

Night-time activity concerns and regulatory urban policies

The development of night-time economies is desirable for cities over the world since they foster the attractiveness of cities and consequently increase tourism's contribution to economic growth. Indeed, the development of an attractive night-time tourism offer can increase a city's competitiveness in relation to other cities with limited (or no) night-time offerings.

As we have seen previously, night-time activities are varied and include a large array of experiences, some more innovative than others. However, traditionally night-time tourism has been associated with 'undesirable' nightlife activity based on binge drinking, noise and anti-social behaviour, or other more general issues such as urban over-development (Eldridge, 2019) and gentrification (Hae, 2011). Therefore, nightlife has frequently confronted residents' interests, and is often rejected by them. In order to minimise and prevent some of these inconveniences, city councils promote various types of policies or regulations. For instance, New York City Council in 2003 proposed to replace the existing cabaret law with a new law that would require a 'Nightlife Licence', which would regulate nightlife businesses based on noise and traffic volumes (Hae, 2011). In London, the city's Night Time Commission was created for the same purpose (see Case study 2).

Case study 2: London Night Time Commission

In 2018, London's night-time economy employed more than 1.5 million people and contributed to the economy of the city, to the extent that jobs in the night-time industries were growing at a faster rate than those in the wider economy as a whole. Two-thirds of Londoners are regularly active at night, including running household errands such as shopping, or socialising. Allied to this, there had been an increase in restaurants, cafes and takeaways open at night, until the onset of the Coronavirus pandemic. In order to maintain London's night-time economy together with safety, an entity was established to evaluate and regulate these issues.

The Night Time Commission has the aim of developing and helping to recognize an ambitious vision for the life of the city, and those who live in it. Recommendations and actions are based on activities that are developed in or for night-time, specifically between 6pm and 6am. In order to maintain London's night-time economy, the Commission provides a report with policy recommendations to the Mayor of London based on "examples of best practice" and "key performance indicators".

The Commission considers opportunities to make the most of shops and public buildings, which are often empty at night, to cope with the economic decline of some streets and localities. For instance, the Commission proposed to use available and appropriate spaces at night, after 6pm, in order to maintain London as one of the World's most vibrant and attractive cities.

The report[1] published by the Commission indicates 10 recommendations to the Mayor. Some examples of those recommendations are:

☐ To set up a night-time data hub including data on the economy, transport, licensing, infrastructure, safety and health, to aid the city's boroughs;

☐ To establish new partnerships across the capital to make London at night more welcoming;

☐ To set up a late-night transport working group to ensure night workers can get to and from work quickly and safely.

This was seen as an opportunity for the city to discover its night side and to become an advanced night-time city compared with the rest of the world.

[1] The full report can be downloaded at: www.london.gov.uk/sites/default/files/think_night_-_londons_neighbourhoods_from_6pm_to_6am.pdf

One of the key issues mentioned above is lighting and its value to transform and beautify the urban space, including either permanent, seasonal or ephemeral illumination. In this sense, Mantei (2012) suggests an appropriate and strategically thought-out policy for lighting, considering all the various lighting types. At the same time, the author advises about the possible risks of illumination such as light pollution, the risk of trivialization of seasonal or ephemeral lighting, a lack of coherence between permanent and ephemeral lighting, "mise en lumière" rejection by the residents and chaotic lighting structure, if management and planning are not centralized and coordinated.

In terms of destination image, the recognition and understanding of the value of lighting is essential (Camprubí & Coromina, 2019a). City marketers should be aware that they can model patterns of the tourism image of their city by creating two different 'scapes' based on night-time and daytime perspectives in order to increase the attractiveness of the destination. In the same vein, a destination image strategy should also be a mechanism to minimise the perceived negative image of some nightlife activities and their negative impacts.

In all, the taken-for-granted differences between night and day urban-scapes, and the type of activities that are undertaken in the diurnal and nocturnal daily cycles of time, have to be kept in mind by tourism policy makers. They offer opportunities for creating innovative tourism experiences adapted to the patterns of behaviour and consumption that visitors desire during a 24 hour time-cycle. Moreover, tourism policy-makers must be aware of the role that day and night urban-scapes play for tourist image creation, and must be used advantageously for branding and marketing the destination, avoiding biased or unrepresentative images in their strategies. They must also be aware of the potential tensions between tourist activities and residents' activities both at night and day and establish the right regulatory systems to minimize the negative aspects of these interferences.

Summary

In this chapter we have explored the relevance of nocturnal lighting to the urban tourism experience and shown examples of how it is used to enhance the attractiveness of visitors' experience after night fall. Attention was also given to the 'border zones' of twilight at dawn and at dusk and the 'nightscape' in which natural phenomena such as the Northern and Southern lights or the importance of Dark-sky Reserves as a counterpart to "mise en lumière" are seen as key components of the touristic experience.

Self-assessment questions:

1. What types of tourism products can we find that have the element of natural or artificial light as their main component?
2. Which are the benefits of lighting management in urban spaces?
3. Which are the main constraints/issues associated with artificial lighting?
4. Why is it necessary to develop lighting and night-time management policies?

References

Ashworth, G. & Page, S. J. (2011) Urban tourism research: Recent progress and current paradoxes, *Tourism Management*, **32**(1), 1-15.

Bourgeois, J. (2002) Le monument et sa mise en lumière, *L'Homme et la société*, **3**, 29–49.

Camprubí, R. & Coromina, L. (2019a) The lighting dimension of perceived tourist image: the case of Barcelona, *Current Issues in Tourism*, **22**(19), 2342-2347.

Camprubí, R. & Coromina, L. (2019b) Residents versus visitors at light festivals in cities: The case of Barcelona, *Journal of Policy Research in Tourism, Leisure and Events*, **11**(3), 455-468.

Demers, A. (2010) L'Offre nocturne de la ville de Québec : vers un tourisme de la nuit, *Rabaska Revue d'ethnologie l'Amérique française*, **8**, 43–19.

Edwards, D., Griffin, T. & Hayllar, B. (2008) Urban tourism research: developing an agenda, *Annals of Tourism Research*, **35**(4), 1032-1052.

Eldridge, A. (2019) Strangers in the night: nightlife studies and new urban tourism, *Journal of Policy Research in Tourism, Leisure and Events*, **11**(3), 422-435.

Eldridge, A. & Smith, A. (2019) Tourism and the night: towards a broader understanding of nocturnal city destinations, *Journal of Policy Research in Tourism, Leisure and Events*, **11**(3), 394-406.

Giordano, E. & Ong, C. E. (2017) Light festivals, policy mobilities and urban tourism, *Tourism Geographies*, **19**(5), 699-716.

Hae, L. (2011) Dilemmas of the nightlife fix: Post-industrialisation and the gentrification of nightlife in New York City, *Urban Studies*, **48**(16), 3449-3465.

Jóhannesson, G. T. & Lund, K. A. (2017). Aurora Borealis: Choreographies of darkness and light, *Annals of Tourism Research*, **63**, 183-190.

Mantei, C. (2012) *Concevoir la lumière comme un lavier de dévelopement touristique*, Atout France. https://www.atout-france.fr/publications/concevoir-la-lumiere-comme-un-levier-de-developpement-touristique

Mitura, T., Bury, R., Begeni, P. & Kudzej, I. (2017). Astro-tourism in the area of the Polish-Slovak borderland as an innovative form of rural tourism, *European Journal of Service Management*, **23**, 45-51.

Weaver, D. (2011). Celestial ecotourism: New horizons in nature-based tourism, *Journal of Ecotourism*, **10**(1), 38-45.

Part 2:
Operational Dimensions of Temporality

The chapters in this section of the book turn their attention to temporal implications on the operational aspects of tourism. Tourism embodies an amalgam of commercial activities which are loosely defined as 'sectors' that service the consumer. While each sector such as accommodation operators, airlines, tour operators, cruise operations, visitor attractions, destination activities, local transport and so forth is affected by temporal considerations, there are a number of common themes, irrespective of the focus or scale of the business operation. This provides the rationale for the structure and content of this section of the book. Examples from various touristic service operations apply the principles within the chapters.

We start with pricing, a core activity for any touristic operation. In Chapter 7, the spatio-temporal basis on which tourism exists provides the rationale for the development of temporal based pricing. The chapter explores how 'temporal pricing' has emerged, its history and development, with particular reference to two of the most important sectors in tourism: hotels and airlines. To understand how temporal pricing emerged, an understanding of the characteristics of the tourism 'industry' are discussed. The chapter then proceeds to examine the process of temporal pricing, built around the three elements of demand forecasting, price discrimination between customer types and characteristics and capacity control.

Building upon this, Chapter 8 introduces key metrics used by tourism operators to improve the performance – and in particular the temporal spread – of their business. The chapter introduces readers to volume and value dimensions of temporal performance, with the worked example of hotel performance metrics and a comparison of airline and hotel indicators. The chapter finishes with a broader application of temporal data sources at national and international levels.

Chapter 9 explores the wider importance for tourism businesses of planning for seasonal variance. The chapter assesses the challenges that seasonal planning presents to a tourism business and goes on to examine the main functions of business planning, showing its application to the case of a local scale bicycle-tourism operation in Germany. The chapter summarises how implementing an operations management framework, including capacity and supply chain planning can help tourism businesses succeed in meeting the challenges of temporal variations in demand. Meanwhile, Chapter 10 focuses on labour in the business operation, specifically seasonal employment and the implications and challenges in a seasonal tourism context at destination and individual business levels and on the workforce itself. Empirical studies in British Columbia and Sweden are used to illustrate the issues.

Finally, in Chapter 11 we confront another endemic issue in the functioning of tourism at local levels, that of the lifestyle operator, for whom temporal trading is widespread. The chapter explores the motivations and typologies of lifestyle operators, and through an 'orientations to work' framework, helps us understand the fluidity of time in the relationships between work and life of lifestyle operators.

7 Temporal Pricing in Tourism

Natalie Haynes and David Egan

Learning outcomes

At the end of this chapter you will be able to :

1. Explain the history and development of temporal pricing in the tourism context.
2. Identify the key temporal factors and characteristics of pricing in the tourism industry.
3. Select suitable temporal pricing tactics and strategies to maximise revenue.
4. Anticipate future changes to temporal pricing practice in the tourism industry.

Introduction

This chapter introduces the key theoretical principles of temporal pricing and explains how they are operationalised within the context of the tourism industry. Temporal pricing is a key element within the overall practice of revenue management which can be defined as *"the process of allotting the right capacity to the right customer at the right price, at the right time, so as to maximise revenue,"* (Nair, 2019: 287). Considering temporal pricing in this way requires the critical relationships between time, demand, price, and capacity to be highlighted. This can be further explored by examining the temporal aspects of a given tourist journey and showing how they are factored into the price decisions made by companies in the industry.

When setting a price for their products and services, hotels, airlines and visitor attractions consider a range of key temporal factors to decide on the prices and inventory controls set for certain periods. Companies will consider the forecasted *demand for the arrival date* of the tourist, *the booking window* or in other words, the time between the tourist making the booking and their arrival, the *duration* of their stay, flight or visit and the tourist's individual balance of *price sensitivity versus time-sensitivity*. Companies use a combination of forecasting, price discrimination and inventory management to manage these temporal factors to maximise revenue and profit often using a blend of automated systems, big data analytics and tacit market knowledge (Egan & Haynes, 2019).

History and development of temporal pricing in the tourism context

Over the last five decades temporal pricing has increased in complexity and digitalisation, with the introduction of automated decision-making systems using data analytics (Hormby et al., 2010). It has also increased in scope, with its use expanding into many sectors within the tourism industry. However, the first use of temporal pricing can be attributed to the airline sector. In 1972 Ken Littlewood of British Overseas Airways Corporation (BOAC) established what is now known as 'Littlewood's Rule'. This rule saw the airline sector experiment with the concept of forecasting demand for different fare types to calculate their different anticipated revenue yields. More discounted fares were issued if their expected revenue value exceeded that of the anticipated values of full-fare tickets (Smith et al., 1992). This evolved further as a result of the deregulation of the airline sector in the late 1970s. A consequence of this was reduced government control over routes and fares, which consequently made the sector more attractive to new entrants, particularly budget airlines, such as PEOPLExpress in the USA.

This immediately led to dramatic increases in competition and falling profits for the existing legacy airlines. It was Robert Crandell, CEO at American Airlines who saw temporal pricing strategies, or yield management as he entitled it, as a solution to this problem and pioneered the first yield management computer system, DINAMO (Dynamic Inventory Allocation Modelling Optimizer). American Airlines introduced non-refundable discount fares controlling the number of seats sold at that lower price whilst saving seats for higher-paying, later-booking passengers, thus maximising revenue per flight through achieving an optimum fare mix. American Airlines was able to control the availability of discounted fares through its computer systems so as to not make the business model unprofitable (Yeoman & McMahon-Beattie, 2016). Budget airlines that did not have these yield systems available tended to respond with drastic, unprofitable and unstainable reductions in ticket prices causing severe financial pressures.

Through these temporal pricing strategies American Airlines achieved revenue growth of 14.5% and key budget competitors such as PEOPLExpress went out of business. In the mid-1980s, the CEO of Marriott International, J.W Marriott Jr. met with Robert Crandell of American Airlines. At a time when the hotel sector was facing similar increased competition from budget operators, J.W Marriott Jr. immediately saw the advantages of utilising similar strategies. He realised that the airline and hotel sectors shared the same characteristics of perishability and constrained supply. Although Marriott preferred the term 'revenue management' to 'yield management' (Hormby et al., 2010), it too implemented a computer system that applied targeted discounts to price-sensitive market segments based on demand and variable lengths of stay, whilst keeping rooms available for later booking, less price sensitive customers.

The current century has witnessed the success of these temporal pricing strategies result in expansion into other areas of tourism through further recognition of the shared characteristics of perishability and constrained supply. The most important areas to try to embrace temporal pricing in recent years have been capacity restricted tourism destinations such as national parks (Schwartz et al., 2012) and visitor attractions such as theme parks (Heo & Lee, 2009).

Operationalising temporal pricing to optimise revenue and the future of temporal pricing

Tourism industry characteristics

One of the key reasons J.W Marriott Jr. met with Robert Crandell of American Airlines was that he saw it possible to replicate the success temporal pricing had at American Airlines, given that the hotel and airline sectors shared some common characteristics which underpinned its operationalisation (see Figure 7.1). Whether the tourism product or service is a hotel room, an airline seat, a ticket to a visitor attraction or entry to a tourism destination, the characteristics of perishability, constrained supply, variable demand patterns, segmentable markets and high fixed costs versus low variable costs all apply. First, *perishability* means tourism inventory has very little or no shelf-life. Once that hotel room, airline seat or entrance ticket has gone unsold, that sale is lost for ever. It cannot be sold at another time, in contrast to, for example, the sale of a pair of shoes in a retail store. Second, *constrained supply* means it is very hard for tourism businesses to vary the supply of their products and services to meet short-term changes in customer demand. Third, the tourism industry is also characterised by highly *dynamic demand patterns* driven by seasonality and other factors such as working patterns and even short-term weather changes. In fact, estimating future demand in the tourism industry is complex because it can be influenced by many different environmental factors as well as characteristics personal to the individual consumer. Fourth, tourists are easily grouped in distinct *market segments* based on their relative time-sensitivity and price-sensitivity. Finally, the industry tends to have *high fixed costs*, such as aircraft leases, but low variable costs associated with selling additional inventory within the overall available capacity.

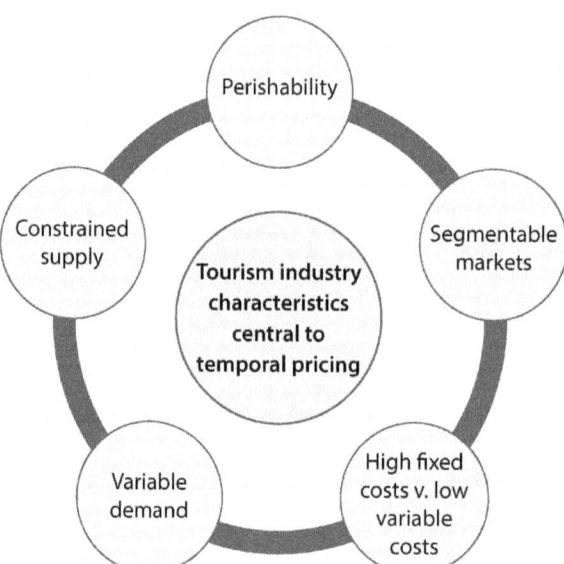

Figure 7.1: Tourism industry characteristics relevant to the temporal pricing process

We will now explain the temporal pricing process that allows for the correct alignment of time, demand, price and capacity to maximise revenue (Yeoman et al., 2001).

Temporal pricing process

The temporal pricing process (as shown in Figure 7.2) is operationalised through the interrelationships between forecasting, price discrimination and capacity control, with the overall aim being to maximise revenue at a specific point in time whilst factoring in the key characteristics of the tourism industry, as highlighted previously.

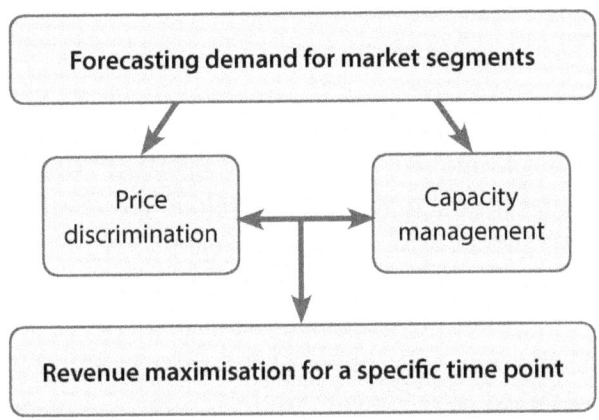

Figure 7.2: The temporal pricing process for revenue maximisation

The first step in this process is the forecasting of demand for market segments. Price discrimination and capacity management requires accurate demand forecasting. A range of factors impact the demand for a certain arrival date in the future and should be factored into the forecasting process. These include weather, customer booking patterns and historical no-show and cancellations rates (Yeoman et al., 2001). It is also common practice for forecasts to be reviewed on at least a daily basis in order to maintain their accuracy and update temporal pricing decisions and strategies as the date of arrival comes closer. This helps to counter the variable demand patterns in the tourism industry, although there may be less variability in areas of the industry that are seasonally driven, such as some tourist destinations and theme parks (Heo & Lee, 2009).

Forecasters must also take into account the differing levels of demand for each market segment at different points in the booking cycle. Forecasters are aware that some market segments will have demand earlier in the booking cycle where the booking window is longer, such as leisure markets, and that some may have greater demand later in the booking cycle where the booking window is short, such as business markets. It is also important to recognise that market segments that are more time sensitive, such as the business market, are generally also less price sensitive, and vice versa for leisure markets. This enables price discrimination to take place where different prices can be charged for the same tourism product or service, based on whether the market segment contains early booking, price sensitive customers or later-booking, price insensitive customers. However, if a tourism business finds itself with lots of spare capacity very close to the date of arrival, the low marginal cost of sales and the high fixed costs, coupled with the perishability of tourism inventory, may mean the business accepts a lower price (Heo & Lee, 2009).

This practice of price discrimination links closely to capacity management and the controls applied to ensure that the right prices go to the right customers at the right time to maximise revenue. The common goal here is to avoid the displacement of higher paying customers who book last minute, by lower paying customers who book early. For some visitor attractions it may also be about protecting the quality of the experience and the sustainability of the destination (Heo & Lee, 2009; Schwartz et al., 2012). Common capacity controls include payment controls such as advanced purchase, cancellation restrictions such as non-refundable fares and duration restrictions such as minimum length of stay. Transparent capacity controls also help justify to customers why they are being charged different prices for the same product or service (Heo & Lee, 2010) and avoid them viewing this as an unfair practice. This is especially important in parts of the tourism industry where temporal pricing practices are less well established and therefore less accepted by customers.

The other element of capacity management is the avoidance of inventory going unsold, especially important given the perishable nature of the tourism industry. It is a long-established reality that customers often either don't show up for their hotel booking, flight or visitor attraction booking or cancel at short notice. This can leave tourism businesses with unsold inventory at the last minute with no hope of recapturing demand that was there earlier in the booking cycle. Therefore, there is a risk that inventory perishes and revenue is lost. Tourism businesses avoid this through the process of overbooking. In essence this means over-selling for a certain arrival date based on a historically based estimate of no-shows and last-minute cancellations. This practice is most common in the hotel and airline sectors. Although over-selling can maximise revenue, it must be managed carefully as if the expected number of no-shows does not occur, other customers may be denied boarding to an aircraft or may be outbooked from a hotel. This can lead to financial and reputational damage for the tourism companies through compensation and customer complaints. A renowned example of this is the United Airlines incident in 2017 where a passenger was forcibly removed from an overbooked flight causing a public outcry on social media (Ma et al., 2019).

However, it is important to recognise that within the area of tourism destination management, the notion of capacity controls often have alternative aims to pure revenue maximisation, especially for some tourism attractions that have a non-profit making business model. In this context capacity controls are more likely to be driven by social and moral issues, such as controlling visitor capacity to protect the destination or attraction from overcrowding and reduce visitors' physical impacts (Schwartz et al., 2012). This is particularly the case for heritage destinations where increasing the entry prices at times of high demand may be used to lessen negative impacts of over-capacity and protect it for future societies (Fyall & Garrod, 1998). The overall aim of temporal pricing processes in these types of attractions would therefore be to manage visitor load rather than increase revenue, in contrast to the more commercially focused areas of the tourist industry such as hotels and airlines. However, there appears to be a gap between the recognised need to use price strategies and capacity control processes in visitor attractions and tourism destinations and their implementation in practice. Here revenue management implementation is often limited in terms of scope and sophistication (Leask et al., 2013). This is currently due to a lack of access to technologically driven revenue management systems, limited use of price discrimination and challenges with the collection of consistent and reliable visitor data.

The future of temporal pricing

Since the revenue management systems designed by American Airlines in the 1970s, temporal pricing processes have been driven by changes and advancements in technology, despite some areas of the industry, in particular visitor attractions, benefitting less from the on-site revenue management systems on offer to hotels and airlines (Leask et al., 2013). One of the most important areas of advancement has been the automation of these revenue management systems (Buhalis & Leung, 2018). However, revenue technologies are likely to continue to develop in sophistication and scope due to advances in technologies such as artificial intelligence, big data analytics and machine learning. It is likely that such developments will also enable an increase in the sensitivity of temporal pricing processes to the needs, values and buying behaviours of individual customers. Prices will not just be set for market segments, but for individuals. The following vignette explores the issues of personalised pricing in more detail.

> ### Vignette: The ethics of personalised pricing
>
> The increasing analytical capabilities of machine learning and artificial intelligence tools embedded within revenue management and pricing systems are allowing for businesses in the tourism industry to provide unique price points for individual consumers based on their consumer profiles and online behaviour (Botta & Wiedemann, 2019). This is often termed 'personalised pricing' and is a practice likely to grow. In fact, Airbnb already has a machine learning tool that creates personalised offers based on 70 different demand and customer related factors.
>
> Instead of tourism businesses having five to ten market segments they will have hundreds of micro-segments based on a range of consumer behaviour factors (Friedll & Hadwick, 2019). For example, for an airline, this could mean an offer in which the travel, wireless connectivity, a checked bag and a warm meal is included in the overall offer price. If the business can add value to the customer's request by addressing an additional need or desire then the result should be increased revenue and customer satisfaction.
>
> However, there are ethical issues that surround the use of personalised pricing which must be considered in order not to result in the opposite reaction – dissatisfied customers and lost revenue! For instance, would you really charge a woman less than a man for entry to a theme park? The answer is hopefully, no! Personalised pricing practices must be based on segmentation using context and behaviour, rather than the protected characteristics of an individual such as gender (Schofield, 2019). This would be viewed as illegal in many countries. In addition, if consumers are unaware of, do not understand, or cannot avoid personalisation, then it can also be viewed as an unfair practice. The key, therefore, to the successful use of personalised pricing is transparency, restricting the use of consumer profiling where possible and allowing consumers the right to opt out of such personalised pricing practices should they wish to (Botta & Wiedemann, 2019).

Summary

This chapter has provided an introduction to the temporal pricing process within the tourism industry through the identification of the key temporal factors and characteristics of pricing in the tourism industry, such as the impact of perishability on pricing. It has explained the history and development of temporal pricing through its origins in the airline and hotel industries to now being considered by visitor attractions and tourist destinations. It has shown how temporal pricing decisions that maximise revenue for tourism businesses can be reached through the accurate forecasting of demand and the setting of price discrimination and capacity management strategies. Finally, we have considered the increasing sophistication of revenue management technologies and the move from pricing based on market segments to personalised pricing through the development of revenue management systems driven by big data analytics, artificial intelligence and machine learning.

Self-reflection questions

1. Can you identify the four key temporal elements of the tourist journey from a revenue management perspective?
2. When might a customer consider price discrimination to be unfair?
3. Can demand forecasting ever be totally accurate?
4. Should automated revenue management systems replace human decision-making on pricing?

References

Botta, M. & Wiedemann, K. (2019) To discriminate or not to discriminate? Personalised pricing in online markets as exploitative abuse of dominance, *European Journal of Law and Economics*, **50**, 381-404.

Buhalis, D. & Leung, R. (2018) Smart hospitality—Interconnectivity and interoperability towards an ecosystem, *International Journal of Hospitality Management*, **71**, 41-50.

Egan, D. & Haynes, N.C. (2019) Manager perceptions of big data reliability in hotel revenue management decision making, *International Journal of Quality & Reliability Management*, **36**(1), 25-39.

Friedll, D. & Hadwick, A. (2019) *Dynamic and Personalised Pricing*, Eye for Travel, UK.

Fyall, A. & Garrod, B. (1998) Heritage tourism: at what price?, *Managing Leisure*, 3(4), 213-228

Heo, C.Y. & Lee, S. (2009) Application of revenue management practices to the theme park industry, *International Journal of Hospitality Management*, **28**, 446-453

Heo, C.Y. & Lee, S. (2010)'Customers' perceptions of demand-driven pricing in revenue management context: comparisons of six tourism and hospitality industries, *International Journal of Revenue Management*, **4**(3/4), 382-402

Hormby, S., Morrison, J., Dave, P., Meyers, M. & Tenca, T. (2010) Marriott international increases revenue by implementing a group price optimizer, *Interfaces*, **40**(1), 45-57

Leask, A., Fyall, A. & Garrod, B. (2013) Managing revenue in Scottish visitor attractions, *Current Issues in Tourism*, **16**(3), 240-265

Ma, J., Tse, Y. K., Wang, X. & Zhang, M. (2019) Examining customer perception and behaviour through social media research - An empirical study of the United Airlines overbooking crisis, *Transportation Research Part E: Logistics and Transportation Review*, **127**, 192-205.

Nair, G.K. (2019) Dynamics of pricing and non-pricing strategies, revenue management performance and competitive advantage in hotel industry, *International Journal of Hospitality Management*, **82**, 287-297.

Schofield, A. (2019) Personalized pricing in the digital era, *Competition Law Journal*, **18**(1), 35-44.

Schwartz, Z., Stewart, W. & Backlund, E.A. (2012) Visitation at capacity-constrained tourism destinations: Exploring revenue management at a national park, *Tourism Management*, **33**, 500-508.

Smith, B.C., Leimkuhler, J.F. & Darrow, R.M. (1992) Yield management at American Airlines, *Interfaces*, **22**(1), 8-31

Yeoman, I., McMahon-Beattie, U. & Sutherland, R. (2001) Leisure revenue management, *Journal of Leisure Property*, **1**(4), 306-317

Yeoman, I.S. & McMahon-Beattie, U. (2016) The turning points of revenue management: a brief history of future evolution, *Journal of Tourism Futures*, **3**(1), 66-72.

8 Measuring Temporal Performance in Tourism

Natalie Haynes and David Egan

Learning outcomes

At the end of this chapter you will be:

1. Aware of the range of data and tools that individual tourism businesses can use to improve their temporal performance.
2. Able to use the metrics to improve individual business temporal performance.
3. Aware of the key metrics used by tourism authorities in understanding the impact of seasonality on tourism at an industry level.

Introduction

This chapter introduces the key metrics used by tourism businesses in improving the temporal performance of their businesses, specifically in the areas of occupancy, revenue and profit. The underlying theory of temporal pricing was explored in detail in the previous chapter. In this chapter we focus on the real-life application of these theoretical concepts and review the range of metrics available to individual tourism businesses. The chapter will also raise some of the real-life trade-offs facing tourism businesses in trying to maximise profits in a dynamic marketplace. In addition we will also consider what these practices mean for individual tourist consumers and the metrics and tools available to consumers to benefit from tourism businesses' desire to maximise the utilisation of their resources. Metrics are a range of quantitative measures commonly used to track and compare performance.

Using the tools of temporal pricing to improve business performance

In this chapter we will explore how two of the key players in the tourism industry (hotels and airlines) use a range of metrics in managing the temporal nature of their products. Both hotel and airline pricing strategies are strongly influenced by the perishability of their product and the seasonal nature of the tourism industry. We will approach the use of the metrics and price setting from the individual or team responsible for ensuring that a company's prices match a customer's willingness to pay. This will vary within each organization but ultimately there will be a person or team responsible for the setting of prices. As noted in the previous chapter, hotels and airlines use what is essentially the same process of trying to maximize revenue in situations of constrained supply and a perishable product where consumers have differing perceptions of value. This then leads to the practice of charging different prices to different customers and to the same customers at different time periods.

Starting with the hotel sector, there is a well-developed set of metrics which are shared and contributed to by individual hotels globally. Although there are several third party providers of these metrics, the best known and most widely available are those provided by Smith Travel Research, usually referred to as STR. In this chapter we will introduce and illustrate how some of the key reports are utilised by hotels in the price decision making and other revenue management techniques. STR is the largest provider of hotel data in the world. Founded in 1985, it aims to provide premium data benchmarking, analytics and marketplace insights for global hospitality sectors. Over 68,000 hotels from 180 countries share their data with STR which is then analysed and presented in the STAR report and other reports for use by the global hotel industry.

The STAR report provides a succinct overview of performance relative to the competitor set. It illustrates a chosen set of competitor hotels to allow a hotel to compare or benchmark its performance to what might be described as its close competitor market, i.e., similar hotels in the local market. It uses the key metrics of Occupancy, ADR (average daily rate), and RevPAR (revenue per available room) for each property and its competitor set as well as Index numbers and Percent Changes. The data displayed covers four points in time: Current Month, Year to Date, Running 3-month (i.e., the average of the current and previous two months), and Running 12-month (i.e. the average of the current and previous 11 months).

The first step is to define and understand how the key metrics are calculated.

- ☐ *Occupancy* is simply 'The number of rooms sold during a specific time period; expressed as a percentage of all rooms available to sell during that same period'

    Total rooms sold/Total rooms available = Occupancy percentage

- ☐ *Average Daily Rate (ADR)* is 'The average (mean) selling price of guest rooms during a specific time period, such as a day, week, month or year'.

    Total Room Revenue/Total Rooms Sold=ADR

- ☐ *Revenue per Available Room (RevPAR)* is the average revenue generated by each available guest room during a specific period. This is often regarded as the key metric as it indicates how successful a hotel is at maximising revenue with the available room resource.

    ADR X Occupancy Percentage = RevPAR

February 2020	Occupancy (%)			ADR			RevPAR		
	My Prop	Comp Set	Index (MPI)	My Prop	Comp Set	Index (ARI)	My Prop	Comp Set	Index (RGI)
Current Month	76.3	77.2	98.8	248.68	230.71	107.8	189.69	178.07	106.5
Year To Date	71.7	72.5	98.8	242.66	225.31	107.7	173.91	163.42	106.4
Running 3 Month	71.7	74.3	96.4	250.44	228.01	109.8	179.44	169.39	105.9
Running 12 Month	82.1	81.8	100.5	302.25	274.15	110.3	248.29	224.16	110.8

February 2020 vs. 2019 Percent Change (%)	Occupancy			ADR			RevPAR		
	My Prop	Comp Set	Index (MPI)	My Prop	Comp Set	Index (ARI)	My Prop	Comp Set	Index (RGI)
Current Month	-2.6	7.6	-9.5	-1.2	-5.3	4.4	-3.8	1.9	-5.5
Year To Date	2.0	2.4	-0.3	-1.6	-4.6	3.2	0.5	-2.3	2.9
Running 3 Month	-0.8	1.0	-1.8	1.0	-3.0	4.1	0.2	-2.0	2.3
Running 12 Month	3.6	2.4	1.2	4.0	7.9	-3.5	7.8	10.4	-2.4

Figure 8.1: Monthly 'performance at a glance': an example of key hotel performance metrics

To see how the RevPAR is calculated, we can use the figures in the table above for the current month.

> ADR X Occupancy Percentage = RevPAR
> 248.68 X 76.3% = 189.69

Looking at how the individual metrics are calculated brings home that maximising the revenue is actually a balancing act between occupancy and room price. Simple economics suggests that if we increase room price this will tend to reduce occupancy and if we want to increase occupancy we need to lower room prices. This is an over-simplification but this logic underlies the balancing act hotel managers and revenue managers undertake every day. Once it is understood that managing hotel room revenue is a balancing act between price and occupancy, then the value of the STAR report becomes obvious as it allows us to review the success of our balancing act compared to our competitor set.

If we use RevPAR as a measure of relative success and refer to Table 8.1, we can see that the hotel that this report refers to is doing relatively well. The RevPAR is higher than the competitor set for all the time periods: e.g., for the current month the RevPAR is 189.69 for our hotel compared to 178.07 for the competitor set. A similar picture can be seen for the other time periods as well. Note the RGI (Revenue Generating Index) figure gives us a quick way of gauging our success: a score of over 100 means that a hotel is doing better than the average of its competitor set. The higher the figure indicates a better performance. Interestingly in this case, although our hotel is doing better than our competitor set, the relative success seems to be under pressure in that the Running 12 month RGI is 110.8, showing a stronger performance than the last Running 3 Month period RGI of 105.9. What is useful is that changes in the competitive position are flagged up for consideration for the hotel management team.

For example, our hotel is doing well overall but the three month RGI figure suggests that the gap in performance with the competitor set is narrowing. The table above indicates that occupancy for the property is doing less well than the competition set. The picture becomes even clearer if we take the second part of the table and consider the percentage changes, where for the current month the hotel is showing a -9.5% decline in occupancy compared to February 2019, causing a fall in RevPAR of -5.5% compared to February 2019. The increase in ADR of 4.4% in 2020 compared to February 2019 looks like it has caused a fall in occupancy and a worse RevPAR compared to February 2019. So pushing up room rates has been counterproductive and reduced the overall RevPAR.

At this point it is worth noting that we have only been considering the summary report. Much more detailed information is provided in the full competitor set report which gives monthly breakdowns over an 18 month period. However, the metrics used are the same though more detailed, enabling an ongoing review of the hotel's competitive position over time. The use of visual data (indices and graphs) allows for easy review of the market trends.

However the STAR reports are not the only metrics that a hotel manager will be using. There will be similar data produced internally at the property level which tends to be reviewed and acted upon several times a day. In addition, most of the major chains have their own data analysis services which supply data on temporal trends and forecasts to aid decision-making. Indeed a major part of the offer for many hotel chains to independent hotel owners to join their chain is the revenue management support systems they offer. For example the IHG group claim their revenue management systems are world beating. Their PERFORM price optimisation system provides daily price recommendations to hotel and revenue managers with a transparent methodology:

"The benchmark rate is an aggregation of competitor prices. By comparing the remaining capacity, remaining demand, competitor rates, and our current BFR [best flexible rate], revenue managers can intuitively judge the reasonableness of the price recommendations. As suggested by the other tabs, price optimization also provides a capability to drill down into demand forecasts, competitive rates, current bookings, and additional pricing analysis"
(Koushik et al., 2012:54).

Airlines use a similar approach, although there is not a central database like STR and the data is individual to the airline. Both industries face similar issues of:

1. the number of seats or rooms available
2. the amount of time remaining to sell the seat or room
3. what competitors are charging for the similar seat or room.

Airline revenue management pricing strategies are based on a sliding scale involving price, inventory, marketing and various sales channels to determine profitable plane ticket prices based on a range of factors such as willingness to buy, competition and destination. Furthermore, the use of algorithms and continuous data collection allows airlines to have true dynamic pricing reacting continuously to changing market conditions, with prices reflecting measures of demand such as number of views of a particular flight or location of an online enquiry.

Table 8.2: summarises and compares the different metrics used by airlines and hotels.

Up to this point we have focussed on the suppliers of hotel and airline services and how they react to variations in demand. However the market is a dynamic system and customers also have a range of metrics available which they will use to try to optimise their own position by trying to get the best value deal from their individual perspective. Indeed, some organisations supply data to both service providers and consumers. Expedia is an example of an online travel agency (OTA) that sells itself as more than a promoter and distribution channel to hotels but also as a major source of data to its suppliers. It also provides the means to potential customers to compare the prices and offers for a significant section of the market at any point in time. OTAs also provide metrics to help customers consider alternatives such as indicators of how busy a particular travel destination is on a particular date.

These OTAs are becoming of increasing importance both to suppliers and customers in the tourism sector. Some of the best known OTAs with global presence include Booking.com, Expedia, Last Minute.com, Trivago, Travel Supermarket, Late Rooms.com, and Hotels.com. Many of these OTAs deal with both accommodation and flights as well as other destination services such as car hire. One of the most popular for flights is Skyscanner whose tag line is that they will:

> "price check with 1,200 travel companies so you don't have to. Sign up for Price Alerts and we'll tell you as soon as fares change on a flight you like"
> (Skyscanner, 2021).

A flight booking site provides a range of options to potential travellers, driven by the use of smart search filters. A customer can search the number of flights, route stops, departure times, available flights by price, seat availability by class of travel etc. In effect customers have a range of metrics to help make a choice that reflects their priorities and is therefore the best value option for them. It is even possible to track prices so that consumers are

Table 8.2

Hotels

Revenue Management	Occupancy	Average Daily Rate (ADR)	Revenue per available room (RevPAR)
The art and science of predicting real-time customer demand and optimizing the price and availability of products to match that demand.	The percentage of available rooms that were sold during a specified period of time. Occupancy is calculated by dividing the number of rooms sold by rooms available. Occupancy = Rooms Sold / Rooms Available.	A measure of the average rate paid for rooms sold; calculated by dividing room revenue by rooms sold.	A metric used to assess how well a hotel has managed their inventory and rates to optimize revenue. Calculated by multiplying occupancy by ADR.

Airlines

Revenue Management	Passenger Composition	Revenue Passenger Kilometres	Yield Unit
This management technique maximizes revenue by enabling the best mix of revenue paying passengers through yield management that involves optimum seat sales in terms of optimum timing and price based on network and fare strategy.	Component ratio of multiple passenger groups including businesses, individual and leisure travellers	Total distance flown by revenue-paying passengers aboard aircraft. Revenue-paying passengers x transport distance (kilometres).	Revenue per revenue-paying passenger per kilometre (or mile). Calculated as revenue ÷ revenue passenger-kilometres.

informed if prices have increased or fallen for their preferred departure date. To illustrate, the authors undertook a search for flights to Hong Kong from Manchester for the month of January 2021. There were 395 options, the cheapest flight being £395 return, the fastest flight being £874. There were only two differences in times, the cheapest flight having a five hour wait in London Heathrow airport as a connection point for the onward flight to Manchester. This information took less than two minutes to obtain.

There are also tools to help travellers identify the cheapest months, when it is best to book in advance and how far in advance. More and more metrics are becoming available and customers are becoming increasingly aware of how to manipulate the temporal variations in demand and price to obtain the best value for themselves. Thus the rapid development of the Internet has provided travellers the opportunity to search and compare travel-related products often via OTAs. At the same time airlines are developing their own online pricing strategies to maximise their revenue (Lee et al., 2020).

Although the range of data provided to customers allows them to act in a way that maximises their value for money, as seen above, there is potential conflict as the primary role of OTAs is to act as a booking agent. Therefore the way the data is presented may be to advantage the booking agent rather than the customer. There has been a recent investigation by the UK's Competition & Markets Authority (CMA) into this issue. The CMA is the UK's official agency that oversees and investigates whether markets are operating fairly.

Vignette: CMA investigation into booking site practices

This investigation examined several practices, including:

- **Search results**: how hotels are ranked after a customer has entered their search requirements, for example to what extent search results are influenced by other factors that may be less relevant to the customer's requirements, such as the amount of commission a hotel pays the site.
- **Pressure selling**: whether claims about how many people are looking at the same room, how many rooms may be left, or how long a price is available, create a false impression of room availability or rush customers into making a booking decision.
- **Discount claims**: whether the discount claims made on sites offer a fair comparison for customers. For example, the claim could be based on a

higher price that was only available for a brief period, or not relevant to the customer's search criteria, for example comparing a higher weekend room rate with the weekday rate for which the customer has searched.

The conclusion the CMA drew was that many of these booking sites were potentially misleading. They published a set of principles in February 2019 to ensure customers are being treated fairly.

Do

- ☐ Be transparent about rankings and 'premium' listings – if the money the business earns affects the search results make it clear for the customer to see.
- ☐ Show customers the total price up front so they are clear on the cost of the purchase and aren't stung by hidden charges.
- ☐ Be honest and tell the whole story if using availability or popularity messaging (for example, "Only 2 rooms left at this price to book on this site", "15 people looking at this hotel for a range of different dates").

Don't

- ☐ Use misleading strike-through pricing (a visual treatment showing a line drawn through the original price and the second lower price offering) and discount claims; discounts must be genuine and compare the same types of rooms for the same stay dates.
- ☐ Hide unavoidable charges, like city taxes and resort fees, until late in the booking process.

(CMA, 2019)

Using seasonal and temporal data for tourism at the national and regional level

National and regional tourism bodies use a range of data to manage their tourism resource where the problems of seasonality and perishability give rise to the need for marketing campaigns to extend the tourism season or create new demands outside the traditional season. There is a wealth of data available in the public domain which is utilised by tourism bodies such as destination management organisations (DMOs) to benchmark and identify competition and market opportunities for their destination areas.

For example the UNWTO (2021a) systematically gathers tourism statistics from countries and territories around the world into a vast database that constitutes the most comprehensive statistical information available on the tourism sector. This database is comprised of over 145 tourism indicators which are updated regularly. The UNWTO introduced the Tourism Data Dashboard (UNTWO, 2021b) which provides statistics and insights for inbound and outbound tourism at the global, regional and national levels. Data covers tourist arrivals, tourism's share of exports and its contribution to GDP, source markets, seasonality and accommodation occupancy.

Data is also produced at a European level by Eurostat (2021), which provides some interesting observations on the seasonal trends in the European Union. For example in 2019, July and August accounted for nearly one third of all annual nights spent in tourist accommodation in the EU. It also shows that more than one in four EU residents' tourism nights in the months of May and September were spent by older people aged 65 or over.

At the country level there tends to be detailed temporal data published by the National Tourism Bodies. In Great Britain for example, a wide range of temporal data is published by Visit Britain/Visit England (2021). In some cases data is available at a regional level such as Yorkshire, in terms of the number of visits, nights and spend by international, domestic and day visitors on a monthly basis. It is sometimes possible to obtain annual data down to a local government level for these metrics.

Data for individual attractions at a destination, regional or national level tends to be more limited. For example, in the UK, the Annual Survey of Visits to Visitor Attractions: Latest results (https://www.visitbritain.org/annual-survey-visits-visitor-attractions-latest-results) contains basic information on the number of visits, prices and marketing expenditure. There are also comparisons with previous years, and trend analysis including sub-national areas such as Yorkshire, while some analysis is available at a local government level.

Summary

This chapter has provided an introduction into how the availability and use of metrics in temporal demand and supply in the tourism sector can be used to guide pricing decisions to maximise revenue for operators. Although the focus has been on the supply side in terms of the detailed use of the metrics as temporal performance indicators, the demand side, from the viewpoint of customers, was also considered. The use of OTAs by customers enables them to maximise their value, in terms of booking flights and accommodation. The chapter concluded with a brief introduction to the range of data available at a global, regional and national levels.

Self-reflection questions

1. Using the data in Figure 8.1 for *'my property for the month of February'* would you recommend the hotel to raise it room rates by 10% if this leads to a fall in occupancy of 5% on room bookings? What arguments would influence your recommendation?
2. Using one of the OTAs, compare the flight costs for a destination of your choice and identify possible reasons why there are different prices on different days of the week for the same journey.
3. Access the UNWTO web site and search for the latest data on tourism arrivals for your home country. Compare how the data has changed from the previous year and identify possible reasons.
4. Search the Visit England site for data on occupancy in the regions of England and compare the Occupancy, ADR and RevPAR for the different regions. Which region has the highest Occupancy, ADR & RevPar for the month of August and how does this compare for the month of January?

References

Competition and Markets Authority, (2019) Accommodation booking sites: how to comply with consumer law, https://competitionandmarkets.blog.gov.uk/2019/02/26/accommodation-booking-sites-how-to-comply-with-consumer-law/

Eurostat (2021) Tourism – Overview, https://ec.europa.eu/eurostat/web/tourism/overview. Accessed 30 June 2021

Koushik, D., Higbie, J. & Eister, C. (2012) Retail price optimization at InterContinental Hotels Group, *Journal of Applied Analytics* **42**, 45-47.

Lee, Y. E., Sun, S. Law, R., & Zhong, L. (2020) Electronic distribution channels of airline ticket(s?). *E-Review of Tourism Research,* **17**, 722-736.

Skyscanner (2021) Why Skyscanner? https://www.skyscanner.net/about-us/why-skyscanner. Accessed 30 January 2021.

Smith Travel Research (2021), STR Hotel Benchmarking. STR Global Ltd. https://str.com/benchmarking. Accessed 30 June 2021.

UNWTO (2021a) 145 World Tourism Statistics. https://www.unwto.org/tourism-statistics-data. Accessed 30 June 2021.

UNWTO (2021b) Tourism Data Dashboard https://www.unwto.org/unwto-tourism-dashboard. Accessed 30 June 2021

Visit Britain/Visit England (2021). England Research and Insights. https://www.visitbritain.org/england-research-insights. Accessed 30 June 2021.

9 Planning for Seasons: the Macro Level

Jana Heimel

Learning outcomes

The chapter will enhance your appreciation of:

1. The foundations of the planning process and systems.
2. The main functions of planning and importance of planning for seasons.
3. Balancing efficiency and effectiveness of planning.
4. The differentiation of long-term (strategic), mid-term (tactical) and short-term (operational) planning.
5. How to develop an operations framework for seasonal planning.

Introduction

The purpose of this chapter is to provoke students to think about the wider implications of temporal variation as challenges for operational planning for tourism businesses. In order to achieve this, this chapter will provide holistic perspectives and case studies from a range of players within the tourism related industries, e.g. tour operators, cruise line operators, transport operators and visitor attractions.

A framework of operational implications of temporal variation is used, including fleet planning, labour and supply chain contracting among others, against which case examples from different sectors are compared and contrasted.

Many tourism businesses such as tour operators and hotels experience seasonal sales volatility over the course of a year. They are extremely susceptible to climatic factors such as winter/summer season variations or weather changes, as well as constructed schedules of human activity (e.g. vacations, bank holiday, events) and thus experience highs and lows in their sales cycle. This volatility can be severe for some types of business (e.g. bike, ski, cruise tour operators and some resort properties). While ski tour businesses and many hotels in high mountains experience a high season in wintertime, bike tour operators and cruise liners are challenged to shift demand through this off-season. In certain cases, it might even be beneficial for them to close the business for this period.

Seasonal sales volatility can cause fundamental cash management issues. During off-season such firms suffer the risk of liquidity squeeze, while in the busy and ramp-up / shoulder season they are likely to generate surplus. In consequence operational management decisions such as closure should be based on a proper financial analysis along with a careful (cash) planning and management analysis (Guilding, 2014). This sales volatility requires a holistic and accurate planning system as part of an operations management framework for steering a business successfully. In this chapter such an operations management framework will be developed and applied to different tourism cases.

Foundations of planning

Planning and the planning system

Planning implies future-oriented thinking, rethinking and determination of goals, measures, means and ways to assure forthcoming target alignment. Planning plays an indispensable role to leverage target-orientation and potential factors of success whilst reducing risk (Horváth & Partners, 2019). Planning provides the creation and protection of a company's future, the achievement of company goals, the promotion of innovations as well as the enhancement of a company's profitability. Its future is safeguarded by anticipating the possible situations so that alternatives are available and a quick response is enabled (Horváth et al., 2020; Fischer et al., 2015).

In general, there is not one single but a range of separate plans for different (functional) departments and/or business units of a company which mainly depend on its underlying steering model. The structured and integrated totality of partial plans and other elements as well as their respective inter-relationships, which should be established and interconnected by consistent principles, form the planning system (Horváth et al., 2020; Fischer et al., 2015).

Seasonal planning challenges

Planning issues for seasons result from seasonal variations which mainly depend on the following two factors: first, climate-determinants such as the four seasons. Besides these periodic-based changes there are also daily variations influencing the sales volume, such as dry and rainy weather which is important to many activity operators such as bike and walking tour organizers or spa places. However, seasons are not as predictable as they used to be a couple of decades ago due to climate change resulting in 'intra-seasonal' variations (Amelung et al., 2007; Koenig-Lewis & Bischoff, 2004). Such intra-seasonal cycles more frequently result from sudden and rather unexpected weather conditions.

Human-determinants such as habits, needs and preferences are another influencing factor of seasonal sales volatility. Vacations and statutory holidays – originally established by humans – are fixed and thus quite well-plannable. However, they still can imply quite a big challenge, especially with regards to staff planning. Since tourism activities on these days depend on weather, market/economic conditions or other unexpected situations as for example political changes or pandemics, this makes it difficult to predict demand and thus plan capacities for those seasons appropriately. As a consequence of such conditions, humans tend to be flexible in their decision making for travel planning. These phenomena challenge operators, depending on staff and other capacities, in their planning and execution of services.

Regarding these seasonal characteristics, planning and budgeting for seasons are becoming as expensive as ever but necessary tasks. As a company increases in size, the formalization of the planning system grows. Figure 9.1 gives an overview of typical challenges leading to a trade-off in planning for seasons with respect to its efficiency and effectiveness (Neely et al., 2001; Jensen, 2001).

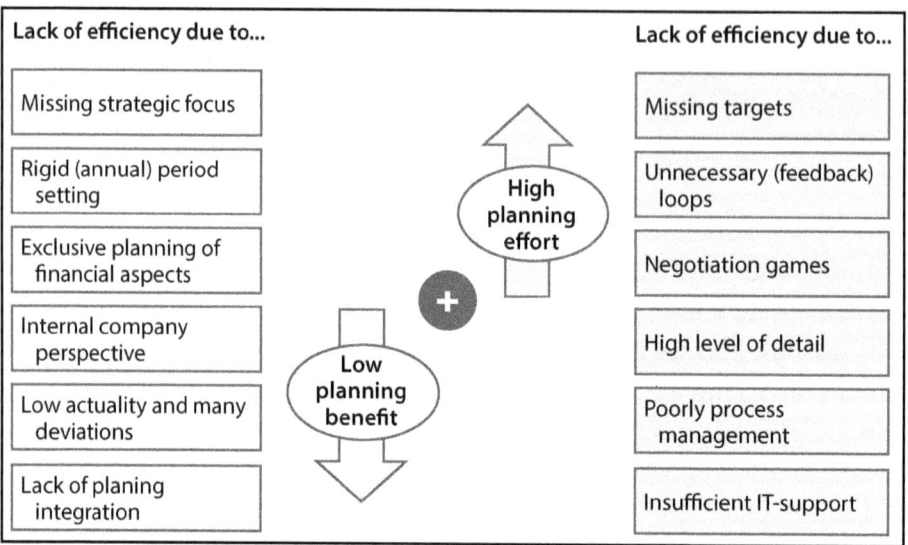

Figure 9.1: Challenges of planning

Functions of planning

The basic functions of planning can be distinguished as follows (Horváth et al., 2020; Pfohl & Stölzle, 1997):

- Coordination
- Performance motivation
- Flexibility
- Safety
- Optimization
- Innovation

Coordination

Planning should contribute to reducing complexity and increasing transparency of the planning objects. Consequently, a horizontal and vertical coordination of all decisions and activities according to the objectives of the organization should be made through the planning process. As part of the coordination function, company goals, macroeconomic development trends and available resources and activities of the individual company areas are ultimately coordinated. A planning-induced increase in efficiency is sought through the coordination of resources and target-based planning. SbB (Stuttgart by Bike), a small tourism business, is used here as a case study to illustrate the planning functions.

> ### Case study: SbB – Stuttgart by Bike
>
> SbB's business model as a start-up business in Germany offering bicycle tours, rentals and courses, is characterized by a four months' off-season starting from November lasting until the end of February. The business is closed for this period and the closure time is used to plan for the new season ahead. Planning starts as late as possible in November to assure reliable data and lower risks in terms of uncertainties. An efficient way of resource allocation is ensured by a standardized year-to-end planning process with an integrated financial liquidity planning. Material resources (bicycles plus equipment) and manpower are planned in November, revised in January, possibly adjusted to new circumstances and introduced to guides in February.
>
> The communication and feedback-loop with SbB staff is used to draft a plan with roughly allocated resources, namely by determining who will guide which tour at what time. In case of a tour cancellation because of rainy weather or travel restrictions, as during pandemic times, guides are allocated other work such as administrative or marketing tasks and remain on the payroll. This allows high long-term planning security for both parties: for guides to earn money and for SbB management to keep on further developing the business. Targets are put into action with the beginning of March.

Performance motivation

Planning should create incentives for staff in order to achieve the desired outcomes and thus positively influence individual performance behaviour. It should therefore provide controllable, challenging, but achievable goals as performance benchmarks. In addition, the possibility of participating in the planning process can be used to actively shape harmonious relationships between employees and managers.

> At SbB, planning data are communicated at the beginning of the new season and continuously reported on in monthly guide meet-ups. In case of significant deviations (as for example, fixed tour dates being shifted by intermediates or tours need to be diverted because of constructions) possible counter measures are introduced to and approved with staff so that they feel and 'own' business targets. Moreover, plans include incentives for employees taking over tours. A monthly award for the best guide of the month is announced to motivate staff to take over tours in order to accomplish company goals.

Flexibility

Planning should predict changes, show the spectrum of possibilities for change and alternative courses of action and thus enable a quick reaction to changes. Flexibility gains in importance with increasing environmental dynamics. For example, climate change increases the chance of warm days during the off-season which may generate unanticipated surges in demand.

> In the case of SbB, it was decided to define 'flexible' in terms of on-demand opening hours (communicated via social media and on the point of sale) for bicycle rentals during off-season. On the one hand it enables the generation of higher sales which in turn helps cover fixed costs, in particular the shop rent, and on the other hand a policy of flexibility helps satisfy demand and consequently increases customer satisfaction. Besides, not having a desk permanently occupied enables SbB to save staff costs in the low- season.

Safety

Planning should identify risks at an early stage, anticipate their effects and initiate appropriate counter-measures. As a result, the security function records the future orientation of the planning and associated uncertainties. In this respect, planning makes a significant contribution to reducing risk.

> SbB rents a retail shop from the city with a two-year limited rental contract but only a one-month cancellation period. Due to this policy, the SbB management integrated scenario planning within the year-end planning. Scenario planning is important in case SbB has to move out of the building. Each year an economic plan was developed to ensure no relocation could occur during the high season, while scenarios with different probabilities of move were examined in the previous year. The challenge was, and still is, to balance the trade-off of not wasting too many resources for planning for a potential move. Yet, the search for new real estate and always having a fall-back option available should still be included in the planning process to assure the sustainable existence of business activities.

Optimization

Planning aims to achieve an optimal selection decision from various courses of action, and thus enforces the quality of the decision. As a basis for the selection decision, operationalized goals must be available, which also include criteria for assessing the degree of goal achievement. By defining future alternative courses of action, there is a reduction in uncertainty.

Finally, SbB must move out of its rented shop. However, it makes use of the opportunity to grow and reopen two more stores, which in turn provides the opportunity to optimize processes and IT-infrastructure.

Innovation

Planning should find new approaches to solutions and strategies in uncertain and difficult times to predict problem situations. In this way, planning should systematically generate previously unknown and neglected alternatives. By creating and securing future room for manoeuvre, there is an increase in flexibility. The Coronavirus pandemic since early 2020 is the ultimate example of this, challenging the global tourism industry to rethink and redefine business models overnight.

> Since SbB was neither allowed to sell bike tours nor events during the Coronavirus shutdown, it started offering emergency repair services so that local tourists and residents could have their broken bikes fixed immediately. This way SbB could compensate lost tour sales with donations for repair services and consequently ensure paying off shop rents. At the same time locals were able to travel and commute in a sustainable and even COVID-19 safe manner during the shut-down.

In corporate practice, the coordination function is most often followed to pursue planning (Nevries et al., 2009).

An operations management framework for seasonal planning

Overview of key success factors

Planning for seasons successfully depends on key factors of success. The following need to be integrated in the planning stages and balanced to plan for seasonal variations appropriately:

- Steering logic and management control
- Value chain management along with
    - ☐ Capacity planning (including staff and other resources)
    - ☐ Supply chain contracting
    - ☐ Building up reputational capital
- Digitization and automation

Figure 9.2 illustrates a best-practice operations management framework integrating the planning stages explained above, supplemented here with the underlying premises and supporting pillars. The critical factors of success are then described, with a focus on implications of temporal variations.

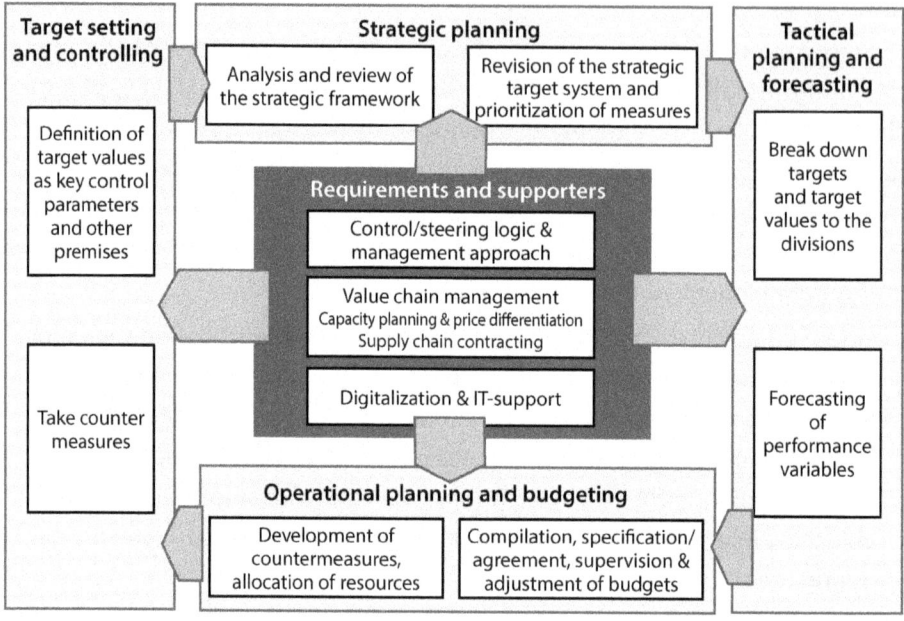

Figure 9.2: The operations planning framework

Steering logic and management control

The overarching management control and steering logic of an organization determine the information needs of its report recipients and thus also the level of detail of the planning (Anthony, 1965). The control/steering concept specifies strategic goals and measures. It comprises processes for influencing behaviour and includes the presentation of key business perspectives (product portfolio, target markets, core competencies) as well as the key success factors in the industry (Anthony et al., 2014; Flamholtz et al., 1985). There are three different control concepts. The starting point for this is the general understanding of the company's claim to leadership (Horváth et al., 2020; Fiedler & Gräf, 2012):

First, in a company run as a **financial holding**, management considers the business areas as pure financial investments. There is no intervention in the operational processes. The management controls the business areas using purely financial indicators such as economic value added (EVA), cash value added (CVA) etc. An example would be a hotel taken over by a capital investor, as has been happening on Mallorca in recent years, as a consequence of the island changing from a low-budget towards an eco-friendly and more luxury oriented tourism destination. The Covid pandemic has even resulted in an increase of sell-offs (Furones, 2020). However, the new investors' interest in retaining the investment might be short-term as typically capital investors seek a two-digit yield and then disinvest within the next three years.

Second, in a company run as a **strategic holding** company, the management controls the business areas through strategic measures of selected value creation functions, using strategic and financial indicators. For instance, imagine that TUI, the world's largest tourism company, becomes part of Warren Buffet's portfolio of stocks. The American billionaire investor, business tycoon, CEO of Berkshire Hathaway (an American multinational conglomerate holding company), usually holds stocks for a long time. Buying shares in the tourism group could be worthwhile in the long term. The Covid-19 travel shutdown caused huge losses in stock for the German travel group. However, since the development of a highly effective corona vaccine, there is great hope in the tourism industry that the travel market could normalize in the medium term. Moreover, TUI is one of the few tourism businesses that received funds from the German government and thus has hope to survive the crisis and in turn contribute to enlarging Buffet's portfolio (Kaletta, 2020; Li, 2020; Partridge, 2020). To satisfy the long-term investor's requirements, the management needs to report on strategic indicators such as market share or customer satisfaction as well as financial indicators.

Third, in a **classic (operational) management** holding company, management is interested in financial, strategic as well as operational aspects and controls the business areas with the corresponding key figures. For example, in a family-owned hotel located in Germany's Black Forest, the management is closely involved in the business areas, controls all strategic and operational business transactions and plans appropriate packages of measures. The forecasting of occupancy and turnover rates, projecting statutory holidays as well as vacation days from visiting countries and monitoring weather forecasts are important to plan for the next season. In this case a holistic understanding of the entire business model and where and how value is being created is indispensable.

To be successful in seasonal planning requires a fundamental operational know-how in terms of market knowledge and good estimation of, for example, how demand and competitors will develop, a realistic estimation of a business' own strengths and weaknesses as well as accurate forecasting of financial developments. In consequence the differentiations of the three forms of management styles dissolve and merge within planning for seasons. The financial, strategic and operative dimension need to be embedded into a three-stage planning framework to guarantee flexibility, in particular when unpredicted events such as external shocks occur and are transferred to all business units of the entire value chain. Accurate and successful seasonal planning thus requires management control along with defined targets to coordinate the different units (e.g. profit vs. cost centres). In consequence, the continuous monitoring and coordination of the entire value chain including main processes is a prerequisite for seasonal planning.

Summary

In this chapter we have first reviewed fundamental knowledge of planning. We considered special issues and functions of planning with focus on seasonal planning challenges. The three main stages of planning: strategic (long-term), tactical/financial (mid-term) and operational (short-term) planning were described. We developed an operations management framework by integrating the three planning stages and identifying key success factors for seasonal planning. We noted the importance of management control and steering logic.

Self-reflection questions for students

With application to any tourism related business or organisation with which you are familiar:

1. What are main functions of planning?
2. What issues need to be considered for planning for seasonal operations?
3. Which three stages of planning can be differentiated?
4. How does a tourism business generate value added services through planning?
5. Which factors are indispensable in planning for seasons?
6. What three types of management control can be differed?

References

Amelung, B., Nicholls, S. & Viner, D. (2007) Implications of global climate change for tourism flows and seasonality, *Journal of Travel Research*, 45 (3), 285-296.

Anthony, R. (1965) *Planning and Control Systems: A Framework for Analysis*, 1st edn, Boston: Division of Research, Graduate School of Business Administration, Harvard University.

Anthony, R., Govindarajan, V. & Hartmann, F. (2014) *Management Control Systems*, 13th edn, Boston: McGraw-Hill.

Fiedler, R. & Gräf, J. (2012) *Einführung in das Controlling - Methoden, Instrumente und IT-Unterstützung*, 3rd edn, Munich: Oldenbourg.

Fischer, T., Möller, K. & Schultze, W. (2015) *Controlling. Grundlagen, Instrumente und Entwicklungsperspektiven*, 2nd edn, Stuttgart: Schäffer-Poeschel.

Flamholtz, E. G., Das, T. K. & Tsui, A. (1985) Toward an integrative framework of organizational control, *Accounting, Organizations and Society*, 10 (1), 35-50.

Furones, J. (2020) Some 100 hotels for sale in Mallorca, https://www.majorcadailybulletin.com/news/local/2020/11/07/74507/mallorca-coronavirus-100-hotels-for-sale-mallorca.html, (Accessed 24 Nov 2020).

Guilding, C. (2014) *Accounting Essentials for Hospitality Managers*, 3rd edn, New York: Routledge.

Horváth & Partners (2019) *The Controlling Concept - Cornerstone of Performance Management*, 1st edn, Munich: Franz Vahlen.

Horváth, P., Gleich, R. & Seiter, M. (2020) *Controlling*, 14th edn, München: Vahlen.

Jensen, M.C. (2001) Corporate budgeting is broken - let's fix it, *Harvard Business Review*, 79 (11), 95-101.

Kaletta, K. (2020) TUI: Geheimer Buffett-Kauf - DER AKTIONÄR hätte einen Vorschlag..., https://www.deraktionaer.de/artikel/aktien/tui-geheimer-buffett-kauf-der-aktionaer-haette-einen-vorschlag-20220901.html , 18 Nov 2020, checked 18 Nov 2020.

Koenig-Lewis, N. & Bischoff, E.E. (2004) Seasonality research: The state of the art, *International Journal of Travel Research*, 7, 201-219.

Li, Y. (2020) Warren Buffett could be building a secret position, online: https://www.cnbc.com/2020/11/17/warren-buffett-could-be-building-a-secret-position.html, checked: 24 Nov 2020.

Neely, A., Sutcliff, M.R. & Heyns, H. R. (2001) *Driving Value Through Strategic Planning and Budgeting*, New York: Accenture.

Nevries P., Strauß, E. & Goertzki, L. (2009) Zentrale Gestaltungsgrößen der operativen Planung, *Controlling & Management*, 53 (4), 237-241.

Partridge, J. (2020) Tui loses £1.8bn so far this year amid Covid-19 travel shutdown, https://www.theguardian.com/business/2020/aug/13/tui-records-1bn-loss-amid-covid-19-travel-shutdown , 13 Aug, checked: 19 Nov 2020.

Pfohl H. & Stölzle W. (1997) *Planung und Kontrolle - Konzeption, Gestaltung, Implementierung*, Munich: Franz Vahlen.

10 Seasonal Employment in Tourism

Tom Baum, Tara Duncan and Deborah Forsyth

Learning outcomes

After reading this chapter, you will:

1. Recognise the relationship between stochastic visitor demand through seasonality and employment in tourism destinations.
2. Recognise the consequences for skills development and the delivery of quality products and services of dependence on a seasonal workforce.
3. Understand the opportunities that seasonal employment offers to mobile workers to gain experience and meet their lifestyle aspirations.

Introduction

This chapter addresses the specific relationship between seasonality in tourism and employment in the sector. Consumer demand for tourism products and services fluctuates hugely in temporal terms across the day, week and season and in response to planned and unexpected events such as economic downturn, natural disasters and, as we witnessed so dramatically in 2020, as a consequence of a global health crisis. This variability is known as *stochastic demand* and has a very significant impact on the lives of those who work in tourism, on who they are, where they come from and what opportunities tourism is able to offer them. In this chapter, we will discuss how the consequences of one particular dimension of stochastic demand, seasonality, impacts on work and the workforce in tourism. We will illustrate what tourism employment in relatively seasonal destinations looks like and how such employment fits into the local community and its economy by drawing on two case examples from the summer destination of Tofino on the west coast of Vancouver Island in British Columbia, Canada and the winter visitor area of Sälen in mid Sweden.

These consequences of stochastic demand cycles on work and the workforce challenge tourism businesses in many parts of the world, especially in rural and coastal destinations. Although there are other forms of variable demand, for example between different service cycles in food service on any one day or relating to mid-week compared to weekend demand for hotel rooms, all of which can impact on the need for labour and skills, here we focus on the most prominent of them, tourism seasonality. Seasonality, which sees tourism businesses in many locations either close down for extended periods during the off-season or cater for very low demand for some or all services, presents one of the apparently insoluble challenges for tourism businesses and their workforce. Seasonal workers are "*a central component of the tourism industry*" (Ooi et al., 2016:246). Tourism businesses require a significant number of workers when there is high demand for their services but the numbers required drops to very low or none at all during the low season or periods when owners either choose or are forced to close completely.

Seasonal demand for tourism products and services has significant consequences for work and employment across a number of dimensions. At one level, these are obvious and very evident when, for example, you visit a beach resort in the middle of winter – limited flights and train services to the destination, shuttered cafes and amusement facilities and empty hotels among other indicators. As a result:

- Employment numbers are significantly reduced during the low season. Closed businesses or those operating at a fraction of capacity require fewer workers in all capacities but particularly frontline roles.
- Many of those who do work in tourism are offered temporary, seasonal contracts which terminate with the end of the operating season or offer reduced, part-time hours during the off-season.

- High season requires the recruitment of workers from outside of the mainstream local labour market – either temporary local workers if available, such as high school or college students or those otherwise unemployed or from other regions or, indeed, other countries.
- Employers are frequently reluctant to invest significantly in the skills and other opportunities of workers who are unlikely be with the business in the long-term. As a result, they may be under-skilled for the work expected.
- Employees lack opportunities for career progression and development in seasonal establishments and must look elsewhere for these.

These consequences of seasonality for the employment of workers can present challenges for tourism businesses seeking to achieve high standards of product and service delivery. This may be difficult for highly branded and standardised operators whose customers have clear brand expectations in terms of both product and service. It may also lead to businesses modifying their product and service offering in line with the skills available in areas such as hotel and restaurant kitchens.

There are also more systemic and less obvious consequences of seasonality for tourism employment which deserve consideration. These include:

- In many seasonal destinations, especially those in peripheral or relatively remote locations, bringing in employees from outside of the locality, frequently from abroad, requires the provision of local accommodation for them. This can be problematic in locations where housing costs are high, supply is limited and prospective employees have to compete with visitors for self-catering accommodation rentals.
- Lack of skills development and indeed, a loss of accumulated skills at the end of the season as in-migrant workers return home means that the cumulative human capital of the destination does not benefit from the employment created by tourism.
- Local workers who depend on seasonal, temporary contracts are disadvantaged when seeking loans for housing and other purposes.

The demands of seasonal tourism businesses for labour do provide working opportunities for those who might find work experience difficult to access. The beneficiaries may include local high school and college students as well as interns from tourism and hospitality programmes in the vicinity or from further afield, including overseas. In some countries, specific arrangements are in place to facilitate temporary workers mobility by migrant workers in order to help meet seasonal skills requirements. One such example is that in place in Canada, where employers and potential workers can choose from a range of temporary programmes depending on criteria such as age, skill level, wage level, family status and mobility or employer sponsorship. The Working Holiday Visa (WHV) is the best-known of the International Experience Canada (IEC) programmes, catering to people under the age of 35 from specific countries who want to experience life and work in Canada for a period of usually up to one year. WHV holders are highly mobile and design their itinerary according to their own interests. Temporary Foreign Worker Programs (TFWP) on the other hand are employer driven, designed to meet labour or skills shortages in specific sectors and/or geographical areas and are organized by high and low wage streams. The controversial Low-Skill Temporary Foreign Worker Program ties migrant workers, often highly educated, to low-skilled jobs such as housekeeping and fast food with a specific employer for usually a period of one to two years.

Seasonal tourism labour markets take on characteristics that further highlight the precariousness of employment that is characteristic of work across the wider tourism sector. We see that assumptions of non-permanence, operating within a defined temporal window and making different workplace demands, notably job flexibility, are of particular importance (Adler & Adler, 2003). Addressing seasonality in hospitality employment especially for remote and economically disadvantaged communities, has long been the focus of government and tourism agency policy (Bohn & De Bernardi, 2020) and has generated a variety of strategies to retain key workers from one season to the next. There have also been product and marketing development initiatives designed to encourage off-season visitation (Fernández-Morales et al., 2016) with some local success stories but limited wider impact. In some contexts, seasonal employment, while not sustainable, offers lifestyle opportunities (Boon, 2006; Lee-Ross, 1999) and possibilities to attract in-comers who may settle permanently in under-populated locations (Möller et al., 2014), introducing fresh ideas and innovation into businesses and communities (Ericsson et al., 2020). Fundamentally, however, seasonality has a negative impact on the sustainability of employment opportunities in hospitality.

Seasonality in practice – the impact on the workforce

Two case examples illustrate how very different destinations accommodate seasonal employment. They are based on the authors' secondary research and, for Case study 1, include data collected through interviews locally.

Case study 1: Tofino, British Columbia, Canada

Tofino is small coastal village located in the Clayoquot Sound UNESCO Biosphere Reserve on the west coast of Vancouver Island, British Columbia, Canada that, according to the 2016 Census, had 1,932 residents. This unique place, where 'everything is connected' lies at the very end of the only paved road to the open Pacific Ocean, surrounded by water on three sides and the Pacific Rim National Park to the south, almost five hours from the city of Victoria. Once a resource-based community focused on fishing and logging, Tofino is now a thriving world-class tourism destination that attracts more than 600,000 visitors annually to enjoy its beautiful beaches, pristine natural environment, world-class surfing and the abundance of recreational opportunities.

Tourism is the lifeblood of Tofino, generating more revenue and more jobs than all other industries combined. Direct tourism alone generates an estimated 2,670 jobs per year, up to 3,600 when considering the indirect and induced spending in the community (Table 10.1). Although an estimated 35% of the total working population of Tofino are employed in direct tourism jobs, the industry depends on thousands of in-migrants each year to sustain itself and the community.

Table 10.1: Annual direct total employment impacts of tourism in Tofino

Impact	Employment		%	All direct tourism jobs	%
Direct tourism	Jobs	FTES	FTES		
Accommodation	1,240	840	49%	Permanent full-time	46
Food & beverage	840	530	31%	Permanent part-time	6
Shopping/retail	200	140	8%	Seasonal full-time	39
Transportation	30	20	1%	Seasonal part-time	9
Outdoor activities & guided tours	270	140	8%		
Other	90	50	3%		
Total direct	2,670	1,720			
Indirect	480	310			
			100%		
Induced	450	290			
Grand total	3,600	2,320			

Source: Economic impact of tourism in Tofino, BC, Final Report by InterVISTAS (4 March 2016) Note: Numbers may not add up due to rounding (FTES= full time equivalents)

Tourism in Tofino is highly seasonal. This and the remote location drive and constrain the temporal flow of tourists and tourism workers. Intrepid storm watchers bolster occupancy during the rainy winter season but the majority of the estimated 1.7 million visitor nights and CA$295 million in tourism revenue occur during peak season from June to the end of September. July and August are at full capacity with occupancy rates around 95% (STR, Inc.) as thousands of tourists and seasonal workers descend on the small community to work and play.

Of the 2,670 direct jobs in tourism, 39% are seasonal full-time and a further 9% seasonal part-time. Eighty percent of all direct tourism jobs are in accommodation or food and beverage, both of which are highly variable. The timing and the temporal nature of these jobs can be an attractive feature to various labour pools who want to experience Tofino, but each group has their own timing and motivation. College students are available during peak season but not in September when occupancy is still at 84%. Lifestyle workers are drawn to the natural amenities but may be more committed to staff housing and surfing conditions than to the employer and the work itself. Labour supply is an ongoing issue as regional and national labour pools are impacted by Canada's aging population and declining birth rate, as the Canadian economy strengthens and fewer people move to British Columbia from other regions of Canada. Labour and skill shortages are not limited to peak season, however. Employers report that chronic labour shortages for housekeepers, cooks, chefs and managers persist, even when the positions are permanent and full-time.

Increasingly, employers turn to international labour markets to meet both their short and longer-term needs. Working Holiday Visa (WHV) holders from Australia, Europe and the UK are drawn to Tofino's natural amenities as part of their Canadian experience and short-term work in hospitality is easily accessible. Several young German workers expressed how exciting it was to be spending time 'at the end of the road' where there was so much to see and do and the surfer lifestyle was so laidback. They worked in housekeeping and the ability to get a job easily and move on when they wanted was important, as was the chance to live in staff housing and feel part of a community in the short time they were in Tofino. Wages are often not a primary motivator but rather a means of subsidizing travel for as long as it takes to experience the destination and move on. For some this is measured in weeks, for others months but this group of WHV holders are more likely to be available during peak and shoulder seasons. Their motivation and mobility is similar to that of domestic lifestyle workers and employers utilize staff housing, flexible schedules and amenities as perks to build community and retain them for the season.

Turnover is high during and after the peak season and attracting and retaining workers to stay throughout the year is problematic. Tofino's remote location, long, rainy winters and lack of amenities such as shopping, entertainment, transportation, fast-food and culture can be a deterrent for potential long-term employees. However the biggest deterrent is a direct result of Tofino's transition from a resource-based economy to one where tourism is the largest economic generator. While tourism contributes an estimated CA$400 million in economic output, the benefits are not evenly distributed and the small number of tax paying residents and businesses are left to bear the cost of the increasing infrastructure costs required to sustain the high volume of visitors and workers during high season. The destination's popularity and remote location drive up daily living costs as well as severely restrict access to affordable housing while low wages associated with tourism employment have depressed the median annual income in the community (Clayoquot Biosphere Trust, 2018). A living wage in Tofino is CA$20.11 per hour, more than CA$8 per hour above the minimum wage that most entry level workers in the tourism industry receive, resulting in many entry level tourism workers needing to work two or three jobs in order to sustain themselves and diminishing the likelihood of ever owning a home. The transition to a tourism based economy and the large number of residents involved in tourism work contributes to the downward pressure on wages towards poverty and challenges the community's ability to meet the United Nations Sustainable Development Goals.

The transient nature, quality and size of the seasonal workforce can have a negative impact on service quality, particularly for larger, higher-end resort and restaurant operations. In 2019 the average daily room rate of more than CA$450 per night during peak season was almost double that of the national average and speaks to the quality of service guests expect.

In Tofino, a number of employers turned to the Temporary Foreign Worker Program (TFWP) as a panacea to high turnover, chronic labour shortages and worker quality. Unlike the WHV program, TFWs sign two-year contracts with a specific employer for a specific job who in turn provides them with full-time work and low-cost accommodation for the duration of the contract. For employers, this 'guaranteed' or 'immobilized' workforce provides stability and flexibility, often at the lowest echelons of the occupational hierarchy in difficult to fill positions, such as housekeeping and the kitchen. Employers involved in the program spoke of the strong work ethic and dependability of these workers and the positive impact their performance had on other workers. During low season in Tofino, occupancy is predominantly on the weekends, which can challenge an employer's ability to provide the requisite full-time hours. This prevented some employers from participating in the program and for those who did, the TFWs often became the 'core' workforce and students, WHV and lifestyle workers filled in the 'gaps'.

Workers from the Philippines and Mexico reported to the author that a contract in Tofino was their only opportunity to access the Canadian labour market, in what they hoped would be the first step towards permanent residency for themselves and their families. Although highly educated, and having had professional careers, these workers left their careers, families and countries to move to Tofino into entry level jobs because wage differentials between Canada and their home country were significant and their primary focus was building a future for their families. This group is financially motivated and the remote location, low cost staff accommodation and lack of shopping are viewed as opportunities to save more money for the future of their families. The natural amenities of Tofino held little interest whereas an opportunity to work overtime or a second job was highly desirable. In conversations with workers from the Philippines they expressed great pride in their reputation as hard workers and in their ability to forge a better future for their children. Many of these workers in Tofino shared that they were inspired by the high number of workers who had arrived under temporary schemes in Tofino and later successfully achieved permanent residency. While many moved to more urban areas once they achieved permanent residency, those with families tended to stay for some years, often with the same employer.

> **Case study 2: Sälen, Sweden**
>
> Sälen is one of Sweden's largest winter tourism destinations located in the county of Dalarna, about a five-hour drive north of from Stockholm. It attracts over 4.8 million visitor nights per year in 55,000 beds of which about 25,000 are commercial. Sälen, which consists of six large ski resorts, has a small population base and relies on an influx of young seasonal workers to cater to the (mainly domestic) visitors. A new international airport, opened in December, 2019, has the potential to provide even more tourism opportunities and change this community substantially.

The continued growth of the tourism industry is relevant for this rural area as it can provide job creation opportunities and allow an in-migration of entrepreneurs, who can have important multiplier effects, through the influx of young seasonal employees who (it is anticipated) might stay in the community (Duncan et al., 2020). Lifestyle features are important for young seasonal workers as they provide these rural spaces with tourism experiences and infrastructure. Thulemark et al., (2014) show that a large percentage of young, seasonal workers move to Sälen for reasons other than employment suggesting that lifestyle and leisure are two factors motivating their move to this area (Duncan et al., 2020).

During a winter season over 2000 seasonal workers are recruited to meet the needs of the tourism industry in Sälen and seasonal workers often outnumber local residents in high season. These young adults are recruited from other parts of Sweden, often from the same areas as the tourists who visit Sälen, as the local labour market cannot cover the number of jobs available. Seasonal workers move to Sälen for three to five months and generally stay in accommodation arranged by their employer or in accommodation solely for seasonal workers. As Duncan et al. (2020) point out, this means that seasonal workers tend to live, work and play with other like-minded young workers who are in Sälen for the same reasons.

Alongside the seasonal nature of such work, Thulemark (2017) suggests many seasonal workers are members of occupational communities. Strong social bonds within the community and a strong relation between work and private life characterise occupational communities (Lee-Ross, 1999). A place dimension, in this example the resort area of Sälen, is also important. Seasonal workers choose a destination based not only on the available leisure activities (skiing or surfing for instance), but also because of the social attributes that they have heard exist among young seasonal workers in the place (Duncan et al. 2020). The attractiveness of destinations such as Sälen or Tofino can be found in resort destinations globally.

There may also be a longer-term temporal element for young seasonal workers which they gain through their attachment to place from their time in Sälen. The social attributes and rootedness to place experienced by seasonal workers in destinations such as Sälen have the ability to influence their decisions later in life to move to rural areas. As Duncan et al. (2020) suggest, in an imagined future some young seasonal workers envision themselves moving to rural mountain communities rich in amenities where they can live their desired lifestyle.

Conclusions

The case examples in this chapter illustrate some of the major challenges that confront the relationship between seasonality in tourism and employment in the sector. The examples illustrate the challenges for seasonal tourism destinations in meeting their skills requirements during the tourist season but also reconciling local community needs with the demands of a temporary, in-comer population of workers. These challenges relate to practical concerns such as housing but also touch on issues of cultural integration and, indeed, the authenticity of local tourism products and services when they are delivered by 'others' from outside of the host community. These present wicked problems for which there may be no obvious 'solution' but to which all business managers as well as planners with a stake in tourism must be sensitive.

Self-reflection questions

1. What are the main challenges for tourism businesses that rely on seasonal workers?
2. To what extent can seasonal employment in tourism realise lifestyle aspirations?
3. From your reading of the Tofino case study, what are the merits and demerits of the Canadian Working Holiday Visa and Temporary Foreign Worker Programs?

4. What conditions do seasonal workers need to become 'occupational communities' integrated in a local community, as opposed to segregated communities?

References

Adler, P.A. & Adler, P. (2003) Seasonality and flexible labor in resorts: Organizations, employees, and local labor markets. *Sociological Spectrum*, 23(1), 59-89.

Bohn, D. & De Bernardi, C., (2020) A labour regime perspective on workforce formation in Nordic tourism: exploring national tourism policy and strategy documents. In Walmsley, A., Aberg, K., Blinnikka, P and Johannesson, G. (Eds.) *Tourism Employment in Nordic Countries - Trends, Practices and Opportunities* (pp. 349-373). Palgrave Macmillan, Cham.

Boon, B. (2006) When leisure and work are allies: The case of skiers and tourist resort hotels. *Career Development International*, 11(7), 594–608.

Clayoquot Biosphere Trust (2018) Clayoquot Sound Biosphere Region's Vital Signs 2018. https://clayoquotbiosphere.org/files/file/5d6b13d204f33/Vital_Signs_18_web_final.pdf

Duncan, T., Thulemark, M. & Möller, P. (2020) Tourism, seasonality and the attraction of youth. In Lundmark, L., Carson, D.B. and Eimermann, M. (Eds.). *Dipping in to the North* (pp. 373-392) Palgrave Macmillan, Cham.

Ericsson, B.; Overvåg, K. & Möller, C. (2020) Seasonal Workers as Innovation Triggers. In Walmsley, A., Aberg, K., Blinnikka, P & Johannesson, G. (Eds.) *Tourism Employment in Nordic Countries - Trends, Practices and Opportunities* (pp. 235.256). Palgrave Macmillan, Cham.

Fernández-Morales, A., Cisneros-Martínez, J.D. & McCabe, S. (2016). Seasonal concentration of tourism demand: Decomposition analysis and marketing implications. *Tourism Management*, 56, 172-190.

Lee-Ross D. (1999). Seasonal hotel jobs: an occupation and a way of life. *International Journal of Tourism Research* 1,239–253.

Möller, C., Ericsson, B. & Overvåg, K., (2014) Seasonal workers in Swedish and Norwegian ski resorts–potential in-migrants? *Scandinavian Journal of Hospitality and Tourism*, 14(4), 385-402.

Ooi, N., Mair, J. & Laing, J., (2016) The transition from seasonal worker to permanent resident: Social barriers faced within a mountain resort community. *Journal of Travel Research*, 55(2), 246-260.

Thulemark, M. (2017) Community formation and sense of place–seasonal tourism workers in rural Sweden. Population, *Space and Place,* 23(3), e2018.

Thulemark, M., Lundmark, M. & Heldt-Cassel, S. (2014) Tourism employment and creative in-migrants. *Scandinavian Journal of Hospitality and Tourism,* 14(4), 403-421.

11 Temporality and the Lifestyle Operator

Claire Holland

Learning outcomes

This chapter will provide you with:

1. An understanding of how we define lifestyle operators and the different typologies of lifestyle operators.
2. An appreciation of the role of temporality in the work and life of lifestyle operators.
3. An understanding of how lifestyle operator relationships with work can enhance understanding of the interactivity between operator typologies, notions of temporality and subsequent temporal strategies employed.
4. An appreciation of how lifestyle goals, temporality and orientations to work interact to create different forms of lifestyle operators that may change over time.

Introduction

Globalisation and an increasingly fluid world have given rise to a greater focus on lifestyle and identity development. The desire for a better lifestyle has seemingly continued to increase as individuals search for ways to bring greater meaning to their lives (Stone & Stubbs, 2007; Walmsley, 2003). Such shifts in societal dynamics have in turn seen an increase in the integration of work as part of a wider lifestyle choice in which non-work and work activities amalgamate to create an overall assimilated lifestyle (Duncan et al., 2013; Shaw & Williams, 1994). Resulting from this, we now see a range of lifestyle operators, who are making work choices with varying degrees of lifestyle focus. Many such lifestyle operators are seen within tourism and hospitality, attracted by low barriers to entry and flexible work offered by temporal trading. Whilst previous studies of temporality in tourism have largely focused on temporality as a problem (Baum & Lundtorp, 2001), there are now indications of some individuals proactively using temporality to facilitate a desired lifestyle, through having redefined the conventions of delineated 'work' and 'non-work' time.

This chapter will explore typologies of lifestyle operators, the motivations driving these individuals and the varying levels of lifestyle focus utilising Reed's (1997) Orientations to Work Framework. It will also explore the nature of temporal trading in different sectors and the relationship that lifestyle operators have with temporality. A systematic review of the current literature on lifestyle operators and operations will then be undertaken, supported by empirical case histories of British tourism lifestyle operators working in Chamonix, France, whose life histories reveal a range of lifestyle operator typologies, each of whom has a differing relationship with temporality.

Defining lifestyle business operators

The concept of lifestyle operators in tourism and hospitality is a growing phenomenon. The desire for a better lifestyle continues on an upward trajectory, spawned from societal shifts towards individualism and identity formation in which individuals search for ways to bring meaning to their lives (Stone & Stubbs, 2007; Walmsley, 2003). Lifestyle can be defined as *"the way in which people live"* (Oxford Dictionary, 2011) and so it may be a way of life or style of living that reflects the attitudes of a group of people or an individual. Within the realms of this definition lifestyle can include all aspects of life whether work or non-work. Despite this, many believe a good lifestyle to be made up of those things outside of paid work and often centred around caring (Evans, 2001). Yet in the modern world of work it is difficult to separate work and non-work (Donkin, 2010). Therefore, lifestyle operators may be seen as a by-product of the modern world with the search for individualism and fulfilment of dreams resulting in the desire to change a way of life (Oliver

& O'Reilly, 2010). Rather than a traditional profit orientation, these lifestyle business operators are *"motivated to start a business on the basis of lifestyle or personal factors"* (Walker, 2004 p577).

Tourism and hospitality sectors particularly attract those seeking a lifestyle focus, drawn by the flexible attributes these sectors can offer, the low barriers to entry and opportunities to 'be your own boss' (see Benson & O'Reilly, 2009; Morrison et al., 2008; Stone & Stubbs, 2007). Williams et al. (1989) initially observed the phenomenon of lifestyle and quality of life aspirations in small-scale businesses in tourism, arguing that those that are generally motivated by non-economic goals could be titled 'lifestyle entrepreneurs' rather than 'lifestyle operators'.

The term 'entrepreneur' often carries connotations of risk taking, a need for independence and a need for achievement (Bosworth & Farrel, 2011) and in the past entrepreneurship has been studied from a multitude of perspectives but most notably an economic perspective (Ateljevic & Doorne, 2000). Yet entrepreneurship has also been associated with achieving a quality of life, with links to satisfaction with managing one's-self and life goals, physical and emotional progress of self and family, and the betterment of one's community (Marcketti et al., 2006). Morrison et al. (2009, p13) define lifestyle entrepreneurs as those:

> *"…who are likely to be concerned with survival and maintaining sufficient income to ensure that the business provides them and their family with a satisfactory level of funds to enable enjoyment of their chosen lifestyle"*.

Examples of such entrepreneurs have been identified as tourism business owners in Northumberland (Bosworth & Farrel, 2011), surfers in Cornwall (Shaw & Williams, 1994), surfers in Ireland (Marchant & Mottiar, 2001), bed and breakfast owners in Canada (Getz & Petersen, 2005) and tourism business owners in New Zealand (Ateljevic & Doorne, 2000). Key characteristics that can been seen across these groups include non-profit motivation and risk aversion.

It has further been suggested that lifestyle entrepreneurs are individuals who own and operate businesses closely aligned with their personal values, beliefs, interests and passions (Marcketti et al., 2006). Lifestyle entrepreneurship can be correlated to numerous push and pull factors with redundancy, escapism, quality of life and being your own boss being key drivers (Benson & O'Reilly, 2009; Stone & Stubbs, 2007). Individuals may also be driven by the desire to earn a respectable living, find satisfaction in career attainment and achievements and spend quality time with family and friends. Indeed, Claire (2012) proposes that as Generation Y enter the workforce they are less associating entrepreneurship with risk taking and more with an opportunity to balance work and life, contribute to society and pursue their passions. This indicates a more complex situation in which lifestyle operators are not just merely concerned with 'survival' as outlined by Morrison et al. (1999, p13), but are driven by self and family development, betterment of society and an ability to integrate work and 'non-work' passions.

Typologies of lifestyle operators

The motivational factors as discussed above are seen to drive the degree of lifestyle focus, generating subgroups of lifestyle operators and lifestyle entrepreneurs. Some of these emerging from the literature are as follows:

Family orientated operators

Family-owned businesses form a large proportion of tourism and hospitality related businesses with a significant draw of this type of ownership being the ability to influence and control the boundaries and overlaps between work and family environments. Family businesses can be split into two broad types: 'Family first business' where the business serves the needs of the family or 'Business first families' where family ownership / involvement are deemed central to the success of the business (Getz et al., 2004). Common challenges associated with this type of operator are balancing work and family commitments, informality of operations, lack of training, limited growth mindset and succession planning.

Entrepreneurs

The term entrepreneur often carries connotations of risk taking, a need for independence and a need for achievement (Bosworth & Farrel, 2011). Entrepreneurship has often been associated with achieving a quality of life, which has been linked to satisfaction with management and decisions regarding one's-self and life goals, physical and emotional progress of self and family, and the betterment of one's community (Marcketti et al., 2006).

Commercial home operators

Here the family home becomes the business with hospitality and tourism as the core product. Motivators for operating in this way include creating a dual economy of generating income from the family home as well as lifestyle goals. Central to this business concept are the host-guest relationships which can both support lifestyle aspirations as well as hinder them (Lynch, 2005).

Retirees

Although achieving a better lifestyle is a primary reason for business start-up (Oliver & O'Reilly, 2010), for some retirees, particularly early retirees, work plays an essential role in sustaining the desired lifestyle (Walmsley, 2003). Here then, the relationship with work is monetary related and is a means to an end rather than a pivotal part of the lifestyle (Stone & Stubbs, 2007).

Transnational workers

This group of workers provide a counter to the traditional retiree migration dynamic motivated by permanent escapism leading to an economic focus on work with life outside of work the priority (Benson & O'Reilly, 2009). Instead, the suggestion is of a more integrated work and lifestyle approach. These workers can be conceptualised as 'transnational' workers within the hospitality and tourism sectors in which individuals are mobile in terms of work, society, community and country. This group is exemplified in the case of Duncan et al's (2013) 'backpacker workers' who make purposeful choices about place and work.

'Seeker migrants'

Comparable to retiree lifestyle operators, those coined by Adler and Adler (2018) as 'seeker migrants' share a commitment to work which is external to and goes beyond the job role. Their involvement in work is higher with a desire to develop self-identity and in some cases perfect their craft (Reed, 1997). As such, these individuals pursue alternative lifestyles and careers that are predominantly dictated by their lifestyle aspirations. Yet the careers chosen are often also used to develop and express their social identity (Adler & Adler, 2018). An example of such 'seeker migrants' are people who work across the globe, job hopping from season to season, for example working in ski resorts as a way to facilitate their love for outdoor winter pursuits.

Lifestyle operators and temporality

As has been demonstrated previously, temporality can be broadly broken down into periodic and seasonal variation (see Chapter 1). Whereas seasonality in tourism and hospitality is associated with longer time periods and/or patterns, periodic variation is characterized as week-by-week variations, fluctuations according to the day of the week or between weekdays and weekends and even in levels of activity within an operational day (Goulding, 2009, p4). This latter is particularly pertinent to the experience of small-scale independent tourism and hospitality operators.

To date, the literature has predominantly focused on temporality from a demand perspective in which the market dictates the nature of temporality in the business (see Getz et al., 2004). Goulding (2009) suggests that this is an oversimplification of the reality of the relationship between temporality and lifestyle operators and attempted to unpick these relationships by drawing connections between commercial home enterprise lifestyle operator motivations, the values created and the temporal trading behaviour created. Commercial home enterprises are an inherent feature of the hospitality industry, represented (in the UK) by guest houses, 'bed and breakfasts' and commercial home stay establishments in particular (the AirBnB phenomenon).

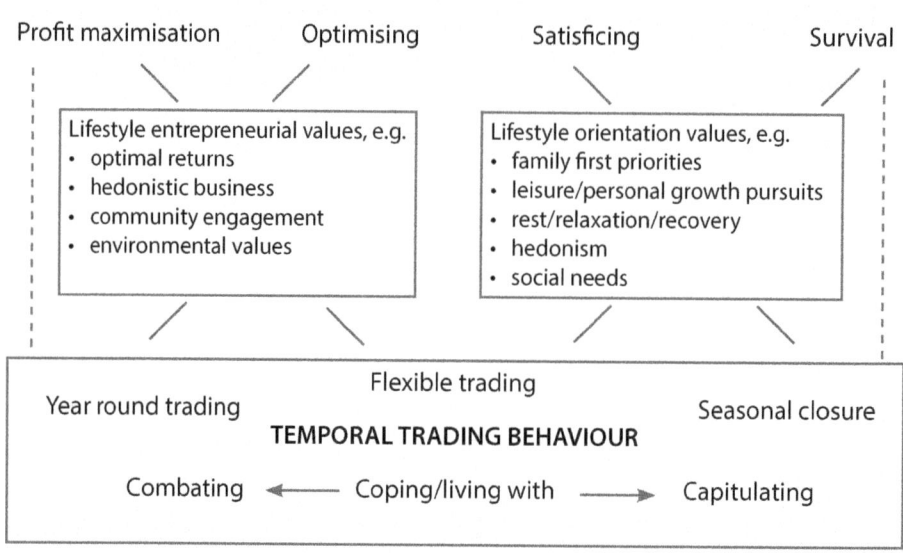

Figure 11.1: Conceptual model of CHE business orientation, lifestyle and seasonal trading (Goulding, 2006)

Expanding upon Getz's (2004) model of how small tourism businesses respond to extreme seasonality of demand (through either 'Coping, Combating or Capitulating' strategies), this model provides a connection between temporality and lifestyle operators. However, with an increasing focus on lifestyle among small tourism business operators and the increase in the range of lifestyle operator typologies, it is prudent to further understand how these interact with the notions of temporality and the subsequent temporal strategies employed. As a model for understanding relationships with work, including whether the business operator's commitment is more external (lifestyle focused) or internal (work focused), Reed's (1997) Orientations to Work Framework can be employed alongside Goulding's conceptual model (Figure 11.1 above) to further unpick lifestyle operator relationships with work and temporality. Reed's (1997) orientations to work framework categorises workers according to whether their locus of commitment is external or internal to the work situation and whether the nature of their involvement with work is as an employment relationship or for social identity.

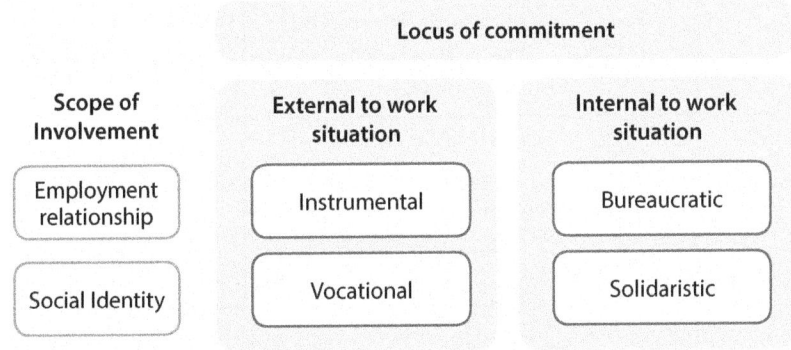

Figure 11.2: The Four Orientations to Work. Source: adapted from Reed (1997)

The four orientations to work of instrumental, bureaucratic, solidaristic and vocational are characterised by two main dimensions. First whether the locus of commitment is external to the work situation, i.e. working to live, or internal to the work situation, i.e. living through work. The second dimension focuses on whether the scope of involvement in work is either as an employment relationship or to build social identity. A summary of each orientation and the typology of lifestyle operator that may be associated with each is as follows:

Table 11.2: Orientations to work and lifestyle operator typologies

Orientation to work	Locus of commitment to work	Scope of involvement with work	Associated Lifestyle Operator Typologies
Instrumental	External	Means to an end. Employment only for financial gain	Family First Businesses Retirees Job hopper Backpacker workers Nomadic workers Expatriates
Solidaristic	Internal	End in itself. Community at work and the work itself is central to building social identity and an opportunity to express oneself	Transnational workers Job hopper Backpacker workers Nomadic workers
Bureaucratic	Internal	End in itself. Employment relationship for career and stability	Business First Family Businesses Career professionals
Vocational	External	Means to an end. Work as a way to build social and self-identity often with a focus on developing craft.	Seeker migrants Lifestyle migrants

Through analysis of the orientations to work and types of lifestyle operators who have to a greater or lesser extent a relationship with work, it is possible to start to draw connections between types of lifestyle operator, their relationship with work and their relationship with temporality. What is important to note is that the work of Reed (1997) introduces the idea of movement between orientations to work over a life course. The next section explores this with the use of some case study examples.

Vignettes of lifestyle tourism workers and operators: Lifestyle applications of Reed's Orientations to Work

We now draw upon a series of short lifestyle operator vignettes gathered as part of a wider study into the relationship that British lifestyle workers in Chamonix, France have with work. These exemplify an instrumental orientation in which lifestyle is the predominant driver. In each case, they exemplify one or more of the 'four orientations' as per Figure 11.2.

Stanley - Lifestyle first

Drawn by a lifestyle of snowboarding, Stanley moved to Chamonix after completing his degree, not having known what to do and having liked it when he visited a friend the previous year. He works purely to fund a lifestyle. For his first season, he worked as a chef with the intentions to work, learn to snowboard and then return to the UK. Still not knowing what to do, he returned to Chamonix and continues to work to fund a lifestyle of living in the mountains.

	Orientation to work	Lifestyle operator typology	Relationship with temporality
Movement over time	Instrumental	Backpacker	Embracing seasonality and periodic variation of temporality as a way to enact desired lifestyle
	Instrumental	Nomadic worker	Embracing seasonality and periodic variation of temporality as a way to enact desired lifestyle

In many respects, Stanley is a 'forever seasonnaire', embracing temporality and using periodic variation in a ski resort during the winter sports season to facilitate his passion of snowboarding in the day and working in the evening.

Janet - Lifestyle first

Janet had always wanted to own a business that would allow her to work in the winter and sail in the summer and so a chalet in a ski resort was a perfect fit. To build their experience, Janet and her husband managed a chalet in Chamonix before approaching investors to set up their own business. During their twenty-year ownership, Janet and her husband were committed to the business but were always driven by lifestyle and the ability to sail. As the family grew and the pressures of work increased, the business was sold and Janet has since set up an online booking service.

	Orientation to work	Lifestyle operator typology	Relationship with temporality
Movement over time	Managing a Chalet -Vocational	Seeker migrant - building skills to set up own business	Embracing seasonality and periodic variation of temporality as a way to enact desired lifestyle
	Own Chalet - Instrumental	Family First Business	Embracing seasonality and periodic variation of temporality as a way to enact desired lifestyle Year-round trading following business success
	Online Booking Service - Instrumental	Family and Lifestyle First	Change in business to online booking service as a way to cope with temporality and regain desired lifestyle

Janet can also be seen to take advantage of periodic variation to meet family needs and non-work interests. Both of the above examples reflect a more traditional lifestyle operator approach with an 'external to work' focus and instrumental orientation to work but an embracing of temporality as a means to achieving a lifestyle.

Yet, we also see that as family needs came to the fore and with the reality of the breaking down of boundaries between work and life, Janet shifts from embracing temporality to taking a strategy of 'capitulation' (in this case, ultimately closing her business). Sherlock (2001) suggests that challenges between the boundaries of public and private, home and work are often inevitable over time and so with many tourism and hospitality businesses being intertwined with personal spaces and place (e.g. the 'commercial home enterprise'), it seems inevitable that work relations and in turn relations with temporality eventually shift. Other examples embrace these notions of temporality but are seen to be driven by an instrumental orientation to work with a focus on family lifestyle needs, akin to Getz et al.'s (2004) family first businesses.

Florence – Family first

As a family-orientated wife, Florence is committed to the business for the lifestyle it affords. An accountant by trade, Florence lived and worked in Leicestershire but when she and her husband started a family, something did not feel right so they moved to Surrey for a better lifestyle. The family worked hard to fund yearly ski holidays and it was on these holidays that Michael became so enamoured with the idea of running their own chalet that they decided to move to Chamonix in the French Alps to set up a business.

	Orientation to work	Lifestyle operator typology	Relationship with temporality
Movement over time	Working in Leicestershire - Bureaucratic	Career Professional	None
	Move to Surrey - Instrumental	Family First Business	Engaging in temporal activities such as ski holidays
	Running own chalet - Instrumental	Family and Lifestyle First	Embracing work and temporality as a means to fulfil lifestyle and family desires

Similarly, those who identify as seasonal, cyclical or transnational workers (such as chalet workers, ski instructors etc...) are also drawn by the temporal nature of tourism with a focus on lifestyle. Their relationship with work can be classified as either instrumental, such as with Stanley in the first example above or 'solidaristic', in which lifestyle satisfaction is gained through involvement with work in which individuals work and 'play' with like-minded individuals:

Doug – Business and community first

From transnational migrant to dedicated host, Doug has worked in many occupations to fund a snowboarding lifestyle, including golf course attendant, tree planter, door to door sales, pub work and sous chef. Since living in Chamonix, he has worked in bars and as a transfer driver before the co-ownership of his current bar with his partner. Doug's original plan was to stay for one last season but he met his partner and ended up staying. He is now driven more by his passion for hospitality and the social community he has built through his work than with the outdoor lifestyle Chamonix offers.

	Orientation to work	Lifestyle operator typology	Relationship with temporality
Movement over time	Multiple jobs to fund lifestyle - Instrumental	Transnational migrant - Job hopper	Embracing seasonality and periodic variation of temporality as a way to enact snowboarding lifestyle
	Passion for hospitality and community - Bureaucratic	First Family Business (community driven)	Coping with temporality

Here Doug has transitioned from being a transnational worker, moving from season to season to engage in his passion of snowboarding in which he embraced temporality, to now employing a strategy of 'coping' with temporality. As the impact of the pressures of temporal trading and elongated seasons have increased and reduced his leisure time, Doug has instead turned to a solidaristic relationship with work in which satisfaction is achieved internally to the work situation through a community of likeminded individuals.

Yet, the developing lifestyle operator literature and instances, such as surfer businesses in Cornwall (see Shaw & Williams, 1994) indicate an alternative perspective in which individuals are displaying greater entrepreneurial attributes such as risk taking, creativity and insight whilst still prioritising a lifestyle (Ateljevic & Doorne, 2000). This could be interpreted as a greater vocational or bureaucratic orientation to work. The lifestyle workers in Chamonix also reflect this trend, with many displaying a vocational orientation in which they are striving to integrate their lifestyle passions with their work.

Jessica and Linda – career minded to lifestyle first

A highly driven job hopper with a passion for the outdoors, driven by skill development and her desire for autonomy, Jessica is keen to develop her work in ways that suits her passions. Jessica has worked in TV production, personal training, has set up a triathlon club and ran a gym in London. She moved to Scotland to work in outdoor education and gain the International Mountain Leader Qualification. She worked freelance in the UK and seasons in Chamonix before meeting her now husband (also a guide) and moving to the Alps to set up an ever-expanding guiding business.

Business partner Linda has worked for the BBC for 28 years as a journalist, presenter and a producer. For 17 years she worked winter and summer seasons in Chamonix for increasingly longer periods with the goal of doing something that she enjoyed and having the lifestyle the location afforded. Driven by this yearning, she gained her International Mountain Leader Qualification, worked as a hiking and snowshoe guide and then set up a mountain activity business with Jessica.

After embracing the lifestyle temporality afforded them, as the work began to feel more like work and less like their passion, the seasonal nature of their work has now become a barrier to being able to achieve the lifestyle they desire. Both have followed similar trajectories in regard to their relationships with work and temporality:

	Orientation to work	Lifestyle operator typology	Relationship with temporality
Movement over time	Multiple jobs - Bureaucratic	Career professional / Job hopper	None
	Move to Scotland / Seasonal work in Chamonix - Vocational	Seeker migrant - Skill building	Embracing work and temporality as a means to fulfil lifestyle desires
	Owning guiding business - Vocational	Lifestyle first migrant	Embracing work and temporality as a means to fulfil lifestyle desires within work as well as outside of work
	Owning guiding business - Instrumental	Lifestyle first migrant	Combating temporality to afford a lifestyle outside of work

Here we still see lifestyle as a central focus but also an optimising approach to the business (see Goulding, 2009). Self-development and self-satisfaction play a role in driving the business forward. The life journeys also suggest that the temporality of tourism has facilitated not only their skill development that led to this point through seasonal job hopping, but also now allows for the time in which they are able to pursue their wider lifestyle passions.

Others have initially displayed a vocational orientation to work and an embracing of temporality but have seen shifts in their relationship with both as changes occurred over the life course.

Lauren – career focused to lifestyle focused

With an eye for a business opportunity and a passion for climbing, Lauren set up mountaineering courses for women in Chamonix. Having lived in London for 20 years working in advertising, she became disillusioned and decided to follow her dream. She bought an apartment and was a circular migrant between London and Chamonix until moving permanently. Lauren initially enjoyed entertaining clients and going up on the mountains but monetary concerns caused in part by seasonality, made this less feasible and she took on an arts teaching job to subsidise the business. The business later ceased trading.

	Orientation to work	Lifestyle operator typology	Relationship with temporality
Movement over time	Advertising job - Bureaucratic	Career professional	None
	Moving between London and Chamonix - instrumental	Nomadic / circular worker	Embracing seasonality and periodic variation of temporality as a way to enact desired lifestyle
	Move to Chamonix full time – Vocational / Solidaristic	Lifestyle first migrant	Coping with temporality as a means to fulfil lifestyle desires within work
	Financial constraints and changing job - instrumental	Lifestyle first migrant	Capitulating to seasonality leading to closure of the business

Conversely, some individuals have initially displayed a bureaucratic orientation, focusing on year-round rather than seasonal trading to ensure profit maximisation, in order to facilitate a deferred lifestyle before shifting to a more instrumental and coping strategy to maintain that lifestyle.

Anthony – career professional to lifestyle focus

As a serial entrepreneur, Anthony has lived in Chamonix for 34 years having moved there at the age of 18 to take a gap year in which he worked in a bar. He never left Chamonix. He set up a tour company that he ran for ten years before selling it. He then set up a transfer company that became the sole provider of transfers for a major budget airline. As the airline grew and demanded a greater slice of his income in commission, Anthony eventually walked away and set up an internet based transfer company. When that company was bought out, Anthony took a step back and now manages a holiday company to have a better work-life balance.

	Orientation to work	Lifestyle operator typology	Relationship with temporality
Movement over time	Multiple jobs - Bureaucratic	Career professional	Combating seasonality and periodic variation of temporality to achieve year-round trading to support lifestyle
	Online transfer company - Instrumental	Expatriate	Coping with temporality to allow for greater focus on lifestyle

Such examples highlight the fluidity in the term 'lifestyle operators' and the ease at which individuals may fluctuate between the various orientations. Likewise these examples help shown how the individuals have differing relations with temporality dependent upon stage of their life and life goals. There is also a temporal dimension to these lifestyle operators themselves, meaning the flexible nature that tourism temporality offers fits with the needs of many lifestyle operators and that where needed the level of lifestyle focus shifts. This in turn determines the relationship that operators have with temporality and the strategies employed. To develop the understanding of lifestyle operator relationships with temporality further, the Orientations to Work framework can be overlaid onto Goulding's (2006) Conceptual Model of CHE Business Orientation, Lifestyle and Seasonal Trading (Figure 11.3).

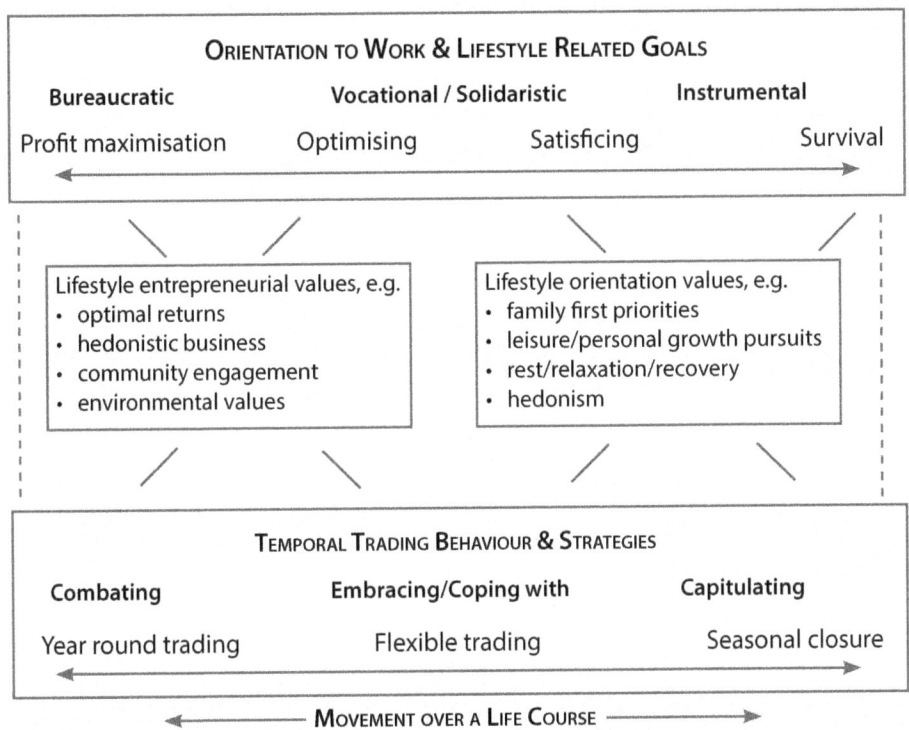

Figure 11.3: Model of Lifestyle Operator Orientation, Lifestyle and Temporality

Summary

In this chapter we have defined and typologised lifestyle tourism operators and introduced conceptual models to illustrate links between temporality and the characteristics of lifestyle and work orientation. We then used case examples of lifestyle operators to apply to these models, demonstrating that the desire for a certain lifestyle remains constant but the form that this takes and the resulting relationship with work differs, dependent on a multitude of factors including life stage.

The role of temporality for lifestyle operators is as either a facilitator or an inhibitor to achieving a desired lifestyle. The individual's/operator's relationship with temporality will also shift as their lifestyle priorities change over the life course. This paints a more complex view of the inter-relationships between lifestyle operators and notions of temporality in tourism, but this helps us conceptualise these links.

Self-assessment questions

1. What are the different types of temporality and how do they impact the tourism and hospitality sectors?
2. There are many different types of lifestyle operator. Summarise the different types and consider the role that temporality plays in each.
3. How can Reed's Orientations to Work framework enhance our understanding of the interactivity between lifestyle operator typologies, notions of temporality and subsequent temporal strategies employed.
4. How and why do lifestyle operator relationships with work and temporality change over time?

References

Adler, P. A., & Adler, P. (2018). Paradise laborers. In *Paradise Laborers*. Cornell University Press.

Ateljevic, I., & Doorne, S. (2000). 'Staying within the fence': Lifestyle entrepreneurship in tourism. *Journal of Sustainable Tourism, 8*(5), 378-392.

Baum, T., & Lundtorp, S. (Eds.). (2001). *Seasonality in Tourism*. Elsevier.

Benson, M., & O'Reilly, K. (2009). Migration and the search for a better way of life: a critical exploration of lifestyle migration. *The Sociological Review, 57*(4), 608-625.

Bosworth, G., & Farrell, H. (2011). Tourism entrepreneurs in Northumberland. *Annals of Tourism Research, 38*(4), 1474-1494.

Claire, L. (2012). Re-storying the entrepreneurial ideal: lifestyle entrepreneurs as hero?. *Tamara: Journal for Critical Organization Inquiry, 10*(1-2).

Donkin, R. (2010). *The Future of Work*. Basingstoke, Palgrave Macmillan.

Duncan, T., Scott, D. G., & Baum, T. (2013). The mobilities of hospitality work: An exploration of issues and debates. *Annals of Tourism Research, 41*, 1-19.

Evans, K. (2001). Relationships between work and life. *The Political Quarterly, 72*, 100-114.

Getz, D., Carlsen, J., & Morrison, A. (2004). *The Family Business in Tourism and Hospitality*. CABI.

Getz, D., & Petersen, T. (2005). Growth and profit-oriented entrepreneurship among family business owners in the tourism and hospitality industry. *International Journal of Hospitality Management, 24*(2), 219-242.

Goulding, P. J. (2006). *Conceptualising Supply-side Seasonality in Tourism: A Study of the Temporal Trading Behaviours of Small Tourism Businesses in Scotland*, Unpublished PhD Thesis: University of Strathclyde.

Goulding, P. J. (2009). Time to trade? Perspectives of temporality in the commercial home enterprise. In Lynch,P. and McIntosh, A.J. (eds.) *Commercial Homes in Tourism* (pp. 122-134). Routledge.

Lynch, P.A. (2005) The commercial home enterprise and host: a United Kingdom perspective, *International Journal of Hospitality Management*, 24, 533-553.

Marcketti, S. B., Niehm, L. S., & Fuloria, R. (2006). An exploratory study of lifestyle entrepreneurship and its relationship to life quality. *Family and Consumer Sciences Research Journal*, 34(3), 241-259.

Marchant, B., & Mottiar, Z. (2011). Understanding lifestyle entrepreneurs and digging beneath the issue of profits: Profiling surf tourism lifestyle entrepreneurs in Ireland. *Tourism Planning & Development*, 8(2), 171-183.

Morrison, A., Carlsen, J., & Weber, P. (2008). Lifestyle oriented small tourism [LOST] firms and tourism destination development. In Richardson, S., Fredline,L., Patiar, A. and Ternel, M. (ed), *Tourism and Hospitality Research, Training and Practice: Where the 'Bloody Hell' Are We?*, Feb 11. Queensland, Australia: Griffith University

Morrison, A., Rimmington, M., & Williams, C. (2009). *Entrepreneurship in the Hospitality, Tourism and Leisure Industries*. Routledge.

Oliver, C., & O'Reilly, K. (2010). A Bourdieusian analysis of class and migration: Habitus and the individualizing process. *Sociology*, 44(1), 49-66.

Oxford English Dictionary (2011). *Oxford English Dictionary*.

Reed, K. (1997). Orientations to work: the cultural conditioning of motivation. *The Australian and New Zealand journal of sociology*, 33(3), 364-386.

Shaw, G., & Williams, A. M. (1994). *Critical Issues in Tourism: a geographical perspective*. Blackwell Publishers.

Sherlock, K. (2001), Revisiting the concept of hosts and guests. *Tourist Studies* 1(3), 271-295

Stone, I., & Stubbs, C. (2007). Enterprising expatriates: lifestyle migration and entrepreneurship in rural southern Europe. *Entrepreneurship and Regional Development*, 19(5), 433-450.

Walker, J.R (2004). *Introduction to Hospitality Management*. Upper Saddle River, N.J, Prentice Hall.

Walmsley, D. J. (2003). Rural tourism: A case of lifestyle-led opportunities. *Australian Geographer, 34*(1), 61-72.

Williams, A. M., Shaw, G., & Greenwood, J. (1989). From tourist to tourism entrepreneur, from consumption to production: evidence from Cornwall, England. *Environment and Planning A, 21*(12), 1639-1653.

Part 3:
Strategic Responses to Temporality

The four chapters in Part 3 of this book take a more strategic overview of aspects of temporality in tourism. The first three assess temporality from a destination lens. In Chapter 12, the authors introduce readers to how destination management organisations (DMOs) deal with temporality in their efforts to attract visitors. It explores relationships that exist between DMOs, space and time. While it can be seen that managing temporal imbalances in visitor flows adds an additional dimension to the task of DMOs, the authors argue that the real challenges arise from meeting the priorities and agendas of the many stakeholders within the destination. The authors thus guide readers through the processes and practicalities of 'bringing in the business' using case examples to illustrate how various DMOs responded to the hiatus of the coronavirus pandemic in their marketing endeavours and how seasonality sits within a DMO's strategic plan.

Events, specifically event tourism strategies, are the focus of Chapter 13. The author explores the relationship between events and festivals within a destination context and goes on to provide a framework of analysis to typologise events at different scales. Event tourism is ideally placed to confront or embrace temporality in a destination, given the nature of events and festivals, their durations and often local community as well as visitor appeal. Accordingly, the author examines the tools of event strategies before focusing on a specific example, the Tramlines music festival in Sheffield, which occupies a key role in the city's events portfolio.

While marketing has been identified in previous chapters, Chapter 14 addresses directly the fundamental relationships between temporal constructs ('the seasons') and marketing communications, through the lens of temporal marketing. It uses case examples of a visitor attraction and seasonal events to examine how seasonal imagery can be used to create a series of visitor experiences. The chapter introduces readers to semiotics and applies semiotic language and image representations of tourism to portray this relationship.

The final chapter revisits planning for seasons, this time from a strategic perspective and one that can apply either to organisations or businesses charged with developing or promoting tourism in the area, or commercial business operations. It focuses in particular on managing supply chains, the value chain, capacity planning and building up 'reputational capital' as four critical strategic planning activities for a tourism business or organisation. The chapter then goes on to assess how digitization of information, data and communication technologies can be employed in temporal planning, building on the case example first encountered in Chapter 9.

12 Temporality: the Destination Management Perspective

Jean Metcalfe and Paul Fallon

Learning outcomes

After reading this chapter, you should be able to:

1. Recognise the multi-dimensional relationships that exist between space (destinations) and time.
2. Analyse DMO interventions to attract visitors from a temporal perspective.
3. Assess the factors which affect the structure, purpose and significance of DMOs.
4. Evaluate the rationale for destination management planning.
5. Recognise the changing priorities and operational approaches of DMOs.

Introduction

This chapter considers temporality from the perspective of the Destination Management (or Destination Marketing) Organization, commonly known as the DMO. DMOs 'do' destination management, which *"essentially equates to management processes that aim to attract visitors"* (Laesser & Beritelli, 2013: 47). As such, DMO activity comprises interventions to develop the visitor economy of a specific place, relating mainly, but not limited to planning, lobbying, marketing and service co-ordination (Laesser & Beritelli, 2013).

Depending upon their individual contexts, DMOs have varying responsibilities but they share the one related to marketing. This issue is considered in more depth, later in the chapter.

Our discussion focuses on the relationships which exist between the DMO, time and space (in this case 'place'). Albert Einstein concluded that time and space represent dimensions within which we think rather than within which we live. Consequently, we are concerned with how and why DMOs operate within these two inter-related paradigms, and especially in their actions for, and reactions to, change. The work of DMOs is complex, fluid and multi-dimensional. Whilst this can be said of many organisations, it is especially significant in destination contexts given their composite nature. Destinations are the sum of many parts, notably including a wide and diverse range of public, private and voluntary stakeholders. Given its association with man-made blending of components, the use of the term 'amalgam' to describe destinations (e.g. Buhalis, 2000), is very appropriate.

An iceberg analogy is helpful to explain this chapter's authors' position further. We argue that the biggest challenge for DMOs is not due to the temporal changes of seasonality and visitor numbers (representing the visible tip of the iceberg, and relating to marketing) but rather the temporal changes in the priorities and agendas of their stakeholders (representing the larger and hidden part of the iceberg, and relating to the other activities mentioned in the opening paragraph). For this reason, the chapter is divided into two parts. The first part, entitled *'Bringing in the Business'*, considers the relationships that exist between DMOs, space and time and refers to interventions to attract visitors. The content may be familiar to you already, but possibly not as seen through a temporal lens presented here.

The second part *'Through a glass, darkly'* explores the external and internal dynamics of DMOs, explaining how and why their agendas may change and the implications of these diversions. Its title is borrowed from St Paul's first epistle to the Corinthians (in the Christian Bible), acknowledging the need to look beyond the surface in order to understand what is really happening. This second perspective is timely given significant changes in the landscape of DMOs – especially at regional and local aggregate levels – which call into question not only their roles, responsibilities and structures but also their very existence. Importantly, the author of this section – herself an experienced practitioner as well as an academic – emphasises that these changes coincide with increased professionalisation in destination marketing activities.

Part 1: 'Bringing in the business'

DMOs act and operate within their spatial contexts, with the main purpose of attracting visitors to them. These contexts can be categorised in various ways, including: at national, regional or local aggregate level; as urban, seaside, rural and island settings; for business, leisure and VFR traffic; and at different life-cycle stages, e.g. developing, developed or in decline (Butler, 1980). Within a destination, different areas are often used for different purposes and may be at different levels of development over time. As such, spatial contexts represent internal opportunities for tourism development irrespective of whether they are already established or not. Metcalfe and Fallon's (2021) retrospective on the use of tourism to re-generate hitherto unattractive places via the concept of 'Tourism in Difficult Areas' (or TiDA) demonstrates how fluid places can be in terms of their usage. The authors emphasise that, in such cases, the place needs to be 're-imagined' not just physically or in the mind of the visitors, but also in the perceptions of relevant internal and external stakeholders over the long term. This principle can also be applied to existing destinations as they move through different stages of their lifecycle, for example those in decline or perhaps clinging on to their former 'glory days'.

Creating change

Spaces are subject to temporal tourism variations which are closely linked to geographical, social and economic dimensions of place. Some places switch from being popular destinations in high summer season to 'ghost towns' in low winter ones. In such cases, DMOs often develop interventions to extend the season or introduce reasons for visitors to come 'out of season'. Blackpool Illuminations, introduced in 1879, is one of the oldest examples of this strategy. The usage of other places changes dramatically during events and festivals, ranging from 'regulars' such as The Notting Hill Carnival (London) and Pride Festivals (in urban areas around the world) to one-off mega-events such as the Olympics. These changes are expected, and DMOs can play a major role in their planning, promotion, delivery and evaluation. They can also be life- and image-changing interventions for the destination, i.e. there is a long-term legacy. Barcelona is often cited as the prime example of a city 'transformed' via the Olympic Games, notwithstanding that its transformation was also due to wider urban regeneration activities (Blanco, 2009). The development of the 'Tour de Yorkshire' as an annual cycling event, following the global coverage of staging the 'Grand Depart' of the Tour de France, is a prime example of opportunism, legacy creation and long term thinking by Welcome To Yorkshire. However, it should also be noted that the legacy from such interventions can represent a mixed blessing, with positive and negative outcomes developing within different lifecycles. Barcelona's 'success story' is now under review due to concerns about over-tourism, especially from local stakeholders. These concerns would probably not have been on anyone's radar a quarter of a century ago. Notwithstanding that, trends and concerns are dynamic and, as the saying goes, 'hindsight is a wonderful thing'!

Reacting to change

Destinations also experience sudden variations which can be both positive and negative. Places can become fashionable almost overnight due to, inter alia: sporting achievement e.g. Leicester City surprisingly winning the English Premiership in 2016; a 'good' or 'bad' news story; an appearance in a book, film or television series. This appeal can have longevity and reach beyond anything that the DMO could achieve by itself:

> *"Film images persist for decades, provide publicity and create identities. The exposure a film gives a city, province or country is an advertisement viewed by potentially millions of people, an audience that could not be reached through specifically targeted tourism promotions."* (Hudson & Ritchie, 2006: 258)

The prime example of this phenomenon was the *Lord of the Rings* film series (for New Zealand tourism) but arguably this is now superceded by the *Game of Thrones* TV series (filmed in various locations, including Northern Ireland, Croatia and Iceland). Consequently, DMOs have increasingly recognised these opportunities and, in addition to marketing activities related to the shows, operate strategically by actively encouraging filming in their area in order to realise these visitors and their associated benefits.

Alternatively, places can suffer sudden crises which have devastating effects on their people and infrastructure – and also their reputation and image. A prime example is New Orleans following the impact of Hurricane Katrina in 2005. Chacko and Marcell (2008) provide very detailed and comprehensive insight into the event, its aftermath and the subsequent recovery implications for the New Orleans Convention & Visitors Bureau in terms of targeting visitors, overcoming negative images and re-positioning the destination. The case emphasises the need to work with both internal and external stakeholders and acknowledge the dynamics of their individual sensitivities over time. It also indicates how the responsibilities of a DMO can change, in this case to include crisis management.

Marketing heritage - 'Stepping back in time'

We should not overlook the fact that many DMOs inherit an historical legacy and their role is primarily to promote it. Countless destinations are attractive because of their history and what remains of it to see, including: Athens and Petra for ancient history; Britain for its regal and industrial heritage; and Nashville and Liverpool for their musical roots. Here priorities are related to preservation and optimising features in promotional strategies so that visitors can enjoy 'step back in time' experiences. The re-imagining and re-packaging of its unique regal history via the 'Ultimate Royal Bucket List' campaign (https://www.visitbritain.com) by VisitBritain adds a clever, contemporary relevance to a well-established product. Related to this, the DMO has also created a feature to exploit the exposure generated by *The Crown* television series, a major international viewing success. In rare cases, an historical opportunity may not be so well-established. The discovery and re-location, in 2012 and 2015 respectively, of the remains of King Richard III in Leicester exposed the city to the world in a unique way. In 2015, and even before the afore-mentioned sporting success which came to the city later as part of a powerful coincidence, *'The Smithsonian'* magazine (2015) promoted Leicester as the 12th of its '25 Great New Places to See'.

Time as a 'nudge'

In addition to the opportunities and challenges created by changes over time related to the destination 'space', DMOs commonly use space and time dimensions in their promotional activities. Place represents arguably the most significant and visible aspect of both offline and online content; after all, 'a picture says a thousand words' – a concept which is highly relevant in a world where quick fixes and first impressions are increasingly important. In addition to the timing of activities – at key points in the calendar year to coincide with particular events and decision-making moments – DMOs often refer to time more subtly in their attempts to 'bring in the business'. These interventions are mainly focused on the targeting of visitor groups and dwell time at the destination and are based on strategic plans and situation analyses.

Targeting 'repeat' visitors, who have been to the destination before and subsequently have an emotional – perhaps nostalgic – connection with that place (Patwardhan et al., 2020), can represent a more effective strategy than trying to attract new ones. Following the New Orleans disaster in 2005, repeaters were specifically targeted through the 'Fall in love with New Orleans all over again' campaign for this reason. As the accompanying vignette indicates, DMOs thinking about 'quick wins' for attracting visitors – if and when the current Covid-19 pandemic is over – may find this concept especially useful. Borrowing the words of a famous 1960s pop song, many destination campaigns, including London, have encouraged extant visitors to extend their stay using the 'Stay just a little bit longer' slogan. Other DMOs attempt to optimise their stop-over status by encouraging tourists in transit to spend time in exploring their destinations; a good example of this is the Hong Kong Tourist Board's (2021) offers of 'The Ultimate Long Weekend' and 'The Ultimate 12-hour layover'.

Vignette 1: Destination marketing during a pandemic

The closure of national borders, international travel restrictions and local lockdown measures, instigated in 2020 and 2021 to combat the spread of the Covid-19 coronavirus, undoubtedly created a major challenge for DMOs. They were faced with determining the right message to send out to existing and target markets during such an unprecedented and unpredictable time. Where there are challenges, there may also be opportunities and this is evident in some of the innovative responses and campaigns. Of interest and relevance to this chapter is the adoption of time within both the product offer and promotional messages. The notion of time passing might also take on more significance for people living through a global crisis, experiencing long periods of relative isolation and facing mortality.

Visit Britain adopted an approach which acted, during lockdown, as a comforting reminder of times past, promoting links to filmed heritage sites related to popular TV drama series, such as Netflix's *The Crown* and giving an opportunity to 'Escape every day at home' (Visit Britain, 2021). St Lucia (n.d.) also offered time away without travel, using Instagram live streaming of yoga practice in front of spectacular scenery with a simple '7 minutes in St. Lucia' message. Time was again evident in a clever strap line used by Southern Delaware Tourism (n.d.) – 'Long Time, No Sea'. This informal phrase conveys a message of reassurance that the place and the welcome remains for whenever it is safe to return.

Switzerland's DMO, perhaps unsurprisingly for a nation renowned for its sophisticated watch and clock-making, refers to time frequently in both the product offer and promotional messages. Presenting 'The Grand Tour of Switzerland' not only tempts the visitor with views of wide open spaces and spectacular scenery but also draws upon times past with its reference to the Grand Tours of the select band of well-heeled travellers of the 17th and 18th Century. Moreover, they remind us to *"Dream now, travel later. Plan now, travel later. You need more time for yourself, time for your dreams. I need more than just a day, I need Switzerland"* (Switzerland Tourism, 2021).

Part 2: 'Through a glass, darkly' – DMOs over time, a practitioner's perspective

Purpose, progress and professionalisation

Tourist areas emerge, develop and decline over time (Butler, 1980) and equally the DMO lifecycle can follow a similar pattern, changing in purpose, scale, and significance. This can be both positive and negative when it comes to broader understanding, acceptance and, ultimately, effectiveness of the DMO's strategy. Heeley (2015:24) proverbially 'hit the nail on the head' when he listed the many titles that have been adopted by such bodies over time – including Tourist Boards, Destination Management or Destination Marketing Organisations and City Marketing Bureaux, observing that *"DMOs typically fail to set out explicitly what their core purpose is, or else do so in rather an unclear manner."* The fact that the terms 'destination management' and 'destination marketing' are used interchangeably, especially in the academic literature, has arguably created a further lack of clarity. Destination management often signifies a broader range of responsibilities – incorporating planning, guardianship and area maintenance – delivered through relevant authorities. Destination marketing, however, implies a focus on targeted activities specifically focused on driving the visitor economy and delivered by a separate entity – essentially the DMO (Pike & Page, 2014; Heeley, 2015).

This separation of function, including the establishment of bespoke organisations to lead, co-ordinate and deliver the vision for the destination, illustrates the shift to a more 'business-like' approach and increasing professionalism over time. In addition, professional associations and membership organisations, such as Destinations International, have raised the profile of the sector and improved quality of DMO provision with guiding principles such as "*community, advocacy, research and education*" (Destinations International, 2021). These industry lead bodies have arguably played a key role in developing and changing the modus operandi for urban destination marketing throughout the world, with many cities keen to follow this lead in order to share in the lucrative, profile-raising international conference and events market. In particular, and largely since the 1980s, the North American visitor and convention bureaux DMO model has been heavily replicated (Heeley, 2015). Prospective clients, i.e. conference/event organisers, benefit from standard practice and terminology, enabling them to directly compare competing cities (due to the familiar presentation and product offer) whilst also recognising their distinctive USPs. Separating leisure and business tourism marketing approaches, whilst retaining destination brand images and messages, also represents a major step forward in facilitating the targeting of specific markets.

Political change and public private partnerships

Time dictates the nature and scope of DMOs in respect of an area's (or nation's) overall economic strategy. This is especially relevant for destinations – notably the afore-mentioned TiDA ones – for which it often takes a long time to realise their ambitions. Without continuity of investment, this is a challenge for such areas with no established visitor profile, and public sector funding remains an important influence here – where there is policy, the pounds follow, so to speak. In the UK, government policy in respect of tourism and the visitor economy has continued to change over the last 30 years. This has not merely been in line with taxation promises and commitments to reductions in public spending, but also in relation to a belief that the private sector might be better skilled, connected and equipped for effective destination marketing. The wide variation in profile and in national, regional and local policy (and thereby access to public funding) is evident across DMOs in the UK and Europe. This variation has resulted in the many different models, and scales, of DMO that we see today – a few private-sector operations, some wholly public funded tourism/city marketing bureaux and many public-private partnerships (PPPs).

PPPs have become widely championed as the ideal operational and financial arrangement to drive and deliver effective destination marketing strategies and campaigns (Heeley, 2015), but this does not guarantee a certain future. In particular, and of potential contemporary relevance due to the Covid 19 pandemic, an over-dependence on membership fees and income generated from professional and consultancy services can prove especially problematic for PPPs at times when the private sector has its own challenges for survival.

Priorities and performance

The Advances in Destination Management (ADM) forum arguably represents the best single sources of relevant, informed and contemporary debates on the issues relating to the management and marketing of tourist destinations. Given participation from highly esteemed and engaged international scholars and practitioners, the emergent themes represent a shared picture of current academic and practical thinking on DMO significance, responsibilities, performance and challenges. The reader is directed to the original papers in order to grasp the full detail and note the contributors to the discussions. However, the fourteen themes from the 2012, 2014 and 2016 meetings (Table 12.1) tell a significant story. Table 12.1 emphasises that, despite an arguably glamourous image, DMOs deal with a myriad of complex macro- and micro-tourism concerns. Furthermore, these concerns are evolving and shifting significantly even within the short term. Consequently, DMOs need to be flexible, forward-thinking and never complacent. They also need to operate effectively within the systems and contexts of the individual destination – there is no 'one size fits all' approach (Metcalfe & Fallon, 2021). DMOs therefore cannot easily copy from each other's successes, which would be rather like 'dancing on a moving carpet'.

Finally, their structures, practices, roles and performance are under constant revision. Increasingly, this is about justifying their existence, and not necessarily with effective measurement techniques. Admittedly, some of these challenges may not be as critical as others for different DMOs, for those operating at different geographic levels and/or with different resources at their disposal. However, the consensus suggests that these are common to many. For example, whilst some DMOs are looking at how to sustain visitor numbers, length of stay and/or expenditure, others are managing over-tourism and ongoing local crises.

Table 12.1: Key themes in destination management, 2012-2016

2012
The definition and delimitation of destination management
Destination marketing and competitiveness
Sustainable destination development and governance
Implications of above for destination management in practice and in research
Source: Laesser and Beritelli (2013)
2014
The definition of 'destination'
The purpose and legitimacy of DMOs
Governance and leadership in destination networks
Destination branding
Sustainability
Source: Reinhold, Laesser & Beritelli (2015)
2016
Relevance of experiences to the destination concept
Destination strategy and resilience
The future of DMOs
Tourism taxation and regulation
Big data and visitor management
Source: Reinhold, Laesser & Beritelli (2018)

Vignette 2: Tackling seasonality

The Isle of Man, situated off the north west coast of England, faces the challenge of managing significant fluctuations in tourism demand, with summer being their peak season. The world famous TT Motorbike Races take place on the small island in May/June annually, attracting around 45,000 visitors over a relatively short period of time (Isle of Man Economic Affairs Cabinet Office, 2018).

Despite the operational and impact costs, major events, such as the TT, can generate additional visitor spend, raise the profile of the host destination and encourage repeat visits (Bowdin et al., 2010; Roult et al., 2020). Additionally, regular sporting events on this scale are anticipated and planned for, with local residents and businesses undoubtedly aware of the temporary nature of related impacts.

> The *Visit Isle of Man Strategic Plan, 2023* acknowledges that seasonality is an issue for the island, highlighting that occupancy levels are *"close to capacity in the main season"* and recognising that productivity falls in the shoulder months, presenting a challenge for local businesses. To this end, the DMOs vision is *"to establish and promote the Isle of Man as a quality, year round, visitor destination for our target audience"* (Visit Isle of Man, 2021).
>
> Product innovations and promotional actions are proposed as ideas to extend the season: including staging events that are not dependent on the weather; running targeted campaigns to promote short breaks with themes related to culture, food or wellness; and facilitating specialist group visits from car and motorsport clubs (Visit Isle of Man, 2021).

Planning process – from policy to practice

As mentioned previously, government policy provides the starting point, giving the rationale, focus and intent behind tourism economic decision making. National and local area strategies should then follow, reflecting core themes as appropriate, but also recognising local priorities and sharing a vision for each destination with aims and objectives to be achieved over a specified period of time. It can be a lengthy and tortuous process to reach an agreed strategy, requiring synergy with local development plans and involving collaborations and consultations with the many stakeholders. Once the vision is agreed, Destination Management Plans (DMP) detail specific projects needed to achieve the strategic aims including time frames, actions, resources and delivery agents. An adopted DMP is designed to ensure that funds and activities are targeted for maximum return, that duplication is reduced and that stakeholders work together towards a common aim (VisitBritain, 2021). Nevertheless, changes in government, and thereby spending priorities, may hinder, delay or divert the initial plan. Destinations International (2015) suggests the following broad structured process for progressing policy:

1. Strategy (establish the vision and value)
2. Business model (set out funding and governance)
3. Lifecycle (based on outcomes of specific projects)
4. Investment partnership (pursue investors)
5. Implementation (DMOs as management with service deliverables)
6. Communication (progress reports)

In addition to policy and funding level adjustments over time, the rapidly changing tourism market, global competition and unexpected events (economic, environmental and societal) require flexible and adaptive DMOs. This is where there may be advantages in having a separate entity – such as a City Marketing Bureau – to focus solely on destination marketing strategy, planning and campaigns. These activities can still fall in line with the broader DMP vision and objectives but, operationally and structurally, they can be more responsive, having an 'ear to the ground' and the benefits of a truly collaborative approach.

People – participation, networking and collaboration

Effective networking with stakeholders and partnership collaborations has become more important to DMOs over time (Bornhorst et al., 2010). This is especially relevant when support is needed to amend strategy, to deviate from planned actions and to communicate new messages. A network model is championed by Destinations International (2015), who cite 'Wonderful Copenhagen' as a prime example; here the DMO operates many networks, each with its own governance, allowing for significant contributions and innovative projects yet still fitting with the overall strategy. Similarly, community support and engagement are ever more significant in the work of DMOs (Destinations International, 2015), indicating a shift over time from pre-occupation with simply 'bringing in the business' (increasing bed nights) to a holistic destination enhancement approach to benefit residents, businesses and visitors. Glasgow, a leading exponent in city branding and with an impressive track record in the international business and leisure events market, exemplifies this community-centred approach. The current *'People Make Glasgow'* brand illustrates the passion and personality of Glaswegians and the city's proud industrial and cultural heritage. Instead of shying away from difficult images from the past and negative impressions that might still be held, the city's DMO confronts this head on. The earlier *'Glasgow's miles better'* campaign was not just a cheerful logo, but a programme of development activities designed to improve the lives of residents and the welcome to visitors. This started a transformation that has continued through staging successful, high profile events from the European Capital of Culture to the Commonwealth Games. Here we see further evidence that there must be a continued commitment and investment over time and that the best exponents of DMOs do not rest on their laurels or work in isolation but must keep focused on strengthening the destination to the benefit of residents and visitors alike.

Summary

Time plays a significant role in the development of destinations in relation to space, image and the agents of delivery, i.e. DMOs. It represents an opportunity, especially as a form of both product and promotion, but it can also represent a threat, with places becoming irrelevant or outdated. Effective destination management incorporates both pro-active and re-active stances, requiring flexibility and creativity alongside commitment over the longer term. Of particular relevance are supportive policies, continuity of investment, and evident professional and collaborative approaches. Whilst 'bringing in the business', i.e. attracting visitors, has arguably been the dominant paradigm, there has been a shift over time to a more holistic approach focused on strengthening the very essence of the destination and actively engaging the resident and business communities.

Reflection questions and activities

1. How can DMO activities align with the destination life-cycle?
2. What are the practical, commercial and moral challenges for marketing a destination which has undergone a recent crisis?
3. What are the advantages and disadvantages of developing a Destination Management Plan?
4. How and why do the roles and responsibilities of DMOs change over time? (Consider using a case study approach to focus the task)
5. Select a relevant area local to you and evaluate its transformation over time for tourism purposes
6. Choose one of the case studies in the Destinations International (2015) *Destination Next, Practice Handbook* and identify the significant lessons to be learnt from it.
https://destinationsinternational.org/sites/default/master/files/pdfs_Dest_Intl_DNEXT_Practice_Handbook.pdf
7. Consider the development of DMO promotional campaigns before, during and after the 2020-21 lockdowns

References

Blanco, I. (2009) Does a 'Barcelona model' really exist? Periods, territories and actors in the process of urban transformation, *Local Government Studies*, **35** (3), 355-369.

Bornhorst, B.R. (2010) Determinants of tourism success for DMOs & destinations: An empirical examination of stakeholders' perspectives, *Tourism Management*, **31**(5), 572–589.

Bowdin, G.A.J., Allen, J., O'Toole, W., Harris, R. and McDonnell, I. (2010) *Events Management* (3rd Ed.), Routledge.

Buhalis, D. (2000) Marketing the competitive destination of the future, *Tourism Management*, **21**(1), 97–116.

Butler, R. (1980) The concept of a tourism area cycle of evolution, *Canadian Geographer*, 24, 5-12.

Chacko, H. & Marcell, M. (2008) Repositioning a tourism destination, *Journal of Travel & Tourism Marketing*, **23** (2-4), 223-235.

Destinations International (2015) *Destination Next, Practice Handbook.* https://destinationsinternational.org (Accessed: 7 January 2021).

Destinations International (2021) *Destinations International 2021 Business Plan.* https://destinationsinternational.org (Accessed: 7 January 2021).

Heeley, J. (2015) *Urban Destination Marketing in Contemporary Europe: Uniting Theory and Practice.* Bristol: Channel View Publications.

Hong Kong Tourist Board (2021) Unique Hong Kong Experiences. https://www.discoverhongkong.com/uk/index.html (Accessed: 10 January 2021).

Hudson S. and Ritchie, J.R.B. (2006) Film tourism and destination marketing: The case of Captain Corelli's Mandolin. *Journal of Vacation Marketing*, **12** (3), 56-268

Isle of Man Government Economic Affairs Cabinet Office (2018) *TT and Festival of Motorcycling Economic Impact Assessment.* https://www.gov.im/media/1361842/2018-05-08-tt-and-festival-of-motorcycling-economic-impact-assessment-part-1.pdf (Accessed 18th April 2021)

Laesser, C. and Beritelli, P. (2013) St. Gallen consensus on destination management, in *Journal of Destination Marketing & Management*, **2** (1), 46-49

Metcalfe, J. and Fallon, P. (2021) Tourism in difficult areas. In Buhalis, D., (ed), *Encyclopedia of Tourism Management and Marketing*. Edward Elgar Publishing, Cheltenham.

Patwardhan V., Ribeiro M.A., Payini V., Woosnam K.M., Mallya J., and Gopalakrishnan P. (2020) Visitors' place attachment and destination loyalty: Examining the roles of emotional solidarity and perceived safety. *Journal of Travel Research*, **59** (1), 3-21.

Pike, S. and Page, S.J. (2014) Destination marketing organisations and destination marketing: A narrative analysis of the literature. *Tourism Management*, 41, 202-227.

Reinhold, S., Laesser, C. and Beritelli, P. (2015) 2014 St. Gallen consensus on destination management, *Journal of Destination Marketing & Management*, **4** (2), 137-142.

Reinhold, S., Laesser, C. and Beritelli, P. (2018) The 2016 St. Gallen consensus on advances in destination management, *Journal of Destination Marketing & Management*, **8** (2), 426-431.

Roult, R., Auger, D. and Lafond, M.-P. (2020), Formula 1, city and tourism: a research theme analyzed on the basis of a systematic literature review, *International Journal of Tourism Cities*, 6(4), 813-830. https://doi.org/10.1108/IJTC-02-2020-0025

Smithsonian Magazine (2015) *The 21st Century Life List: 25 Great New Places to See.* www.smithsonianmag.com/travel/the-21st-century-life-list-180956324 (Accessed: 10 January 2021).

St Lucia Tourism (n.d.). www.stlucia.org/en_UK/ (Accessed: 27 April 2021)

Switzerland Tourism (2021) The Grand Tour of Switzerland. www.myswitzerland.com/en-gb/ (Accessed: 27 April 2021)

Visit Britain (2021) filming locations for The Crown. www.visitbritain.com/gb/en/discover-crown-filming-locations-britain# (Accessed: 27 April 2021).

Visit Isle of Man (2021) Visit Isle of Man Strategic Plan 2023. https://www.visitisleofman.com/dbimgs/1630%20Visit%20IoM%20Strategy%20A4%20JAN20-WEB.pdf (Accessed: 18th April 2021)

Visit Southern Delaware (n.d.). https://visitsoutherndelaware.com/ (Accessed: 27 April 2021).

13 Temporal Event Tourism Strategies

Mark Norman

Learning outcomes

After reading this chapter, you will be able to:

1. Understand the temporal nature of events within a tourism context,
2. Describe the different event typologies in this context,
3. Describe typical event strategies that can be applied to overcome temporal issues in tourism.

Introduction

Event tourism is the practice of attracting tourists to a destination via a structured event programme or portfolio. The case for academic study of event tourism has been made by several authors (Bramwell & Rawding, 1994; Getz, 2008; Getz, 2016). Destinations may use event tourism for several reasons: to overcome seasonality, for broader socio-economic reasons or to achieve competitive advantage. This chapter will explore the different typologies, challenges and strategies associated with event tourism from a temporal dimension.

Richards and Palmer (2010) discuss the concept of cities becoming eventful. While not all destinations are cities, Richards (2017) suggests events are attractive to destinations (in particular, cities) for both economic and social development. Socially, they can celebrate local heritage, culture or history as a way of boosting social cohesion. From an economic perspective, they provide opportunities for image enhancement, income generation, repositioning and job creation (Pugh & Wood, 2004; Richards, 2017).

Destinations can be defined from a number of perspectives. They may be visitor attractions or resorts, geographic entities or perhaps places characterised by a distinct purpose of visit, such as VFR or business tourism. Destinations with distinct patterns of seasonality, such as ski resorts, may seek to become eventful as a means to overcome temporal issues by developing an event program during their off-season. Connell (2015) examined the role of events in mitigating seasonality within specific tourism attractions, suggesting that having an events programme helps destinations maintain hotel occupancy levels and restaurant use, and brings other economic gains across the year.

While some destinations enjoy the benefits of traditional tourism assets such as landscapes, culture, heritage or leisure pursuits, some are located away from such attractions. These destinations may use event tourism to provide a competitiveness level (Crouch, 2010). For example, post-industrial cities may choose an event tourism portfolio to compete with more culturally rich counterparts (Bramwell & Rawding, 1994). For these places, the effects of temporality may vary depending on the source of visitor footfall. Thus events can be used as a means to rebalance demand across a year to provide a consistent competitive offer.

The concept of using festivals and events as tourist attractions would stand as a potential alternative to traditional tourism assets (Connell et al., 2015; Pacione, 2012). In many ways, events have long been a destination's answer to managing the fluctuations of temporality. If festivals and events can become tourist attractions in their own right, they become a powerful tool for destinations to mitigate the effects of temporal fluctuations in their tourism economy.

Event tourists are those people who primarily travel during the time of an event rather than necessarily travelling for the wider appeal of the destination itself. However, research has suggested that the destination's overall attractiveness still plays a part in event motivated travel (Oh & Lee, 2012). Clearly though, there must be a symbiotic relationship between destination image and event image, since both play a role in travellers' decision-making process (Lai, 2016).

Event temporal typologies

Events typologies have developed through ongoing academic research. One of the most recognised typologies is from Getz (1997), who proposes a pyramid of events as depicted shown in Figure 13.1. This structure suggests that as the number of events in each tier grows, the smaller and more local the events become.

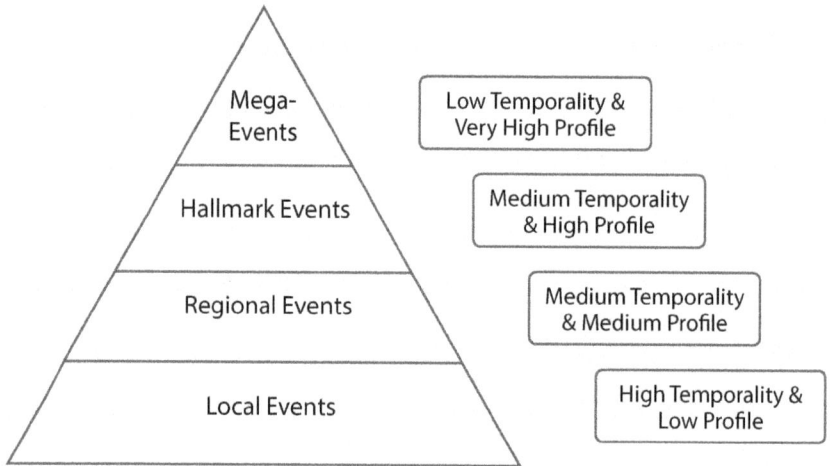

Figure 13.1: Temporality of different events types, adapted from Getz (1997)

At the top of the pyramid are *mega-events* which occur most infrequently but have the most significant potential to attract event tourists. Mega-events typically include major sports events like the Olympic Games or Soccer World Cups that move from destination to destination using a bidding process. Visitors are attracted in large numbers to such events given their high profile. These events are significantly temporal, characterised by a high impact in a concise space of time. For example, the Summer Olympic Games lasts no longer than 16 days while a FIFA World Cup lasts around one month.

As events like these do not occur regularly in a destination, they may have less sustainable impact on the host destination's pattern of seasonality in the years following the event. One of the critical bidding criteria for such events if often their legacy beyond the actual event.

Next are *hallmark events*, which tend to be attached to the host destination. Hallmark events can often be centred around a particular historical or cultural celebration of a destination, implying that moving the event elsewhere would be very problematic. Hallmark events happen periodically, mostly every year, but some occur biannually. For example, the Hong Kong

International Dragon Boat Races takes place each year in Victoria Harbour. The event is annual, highly competitive and often televised. The event itself has a cultural heritage and tradition as part of a public holiday in the city. The event attracts nearly 30,000 spectators which, in itself provide a significant financial injection to the local economy.

Hallmark events will typically have more economic value to destinations looking to overcome temporal challenges, such as boosting the visitor economy at a particular of the year.

Next in the pyramid of scale are *regional events*. These events are smaller in scale than hallmark events and attract a lower level of demand from tourists. There will likely be more regional events spaced out across the year within a destination and a significant proportion tend to be community-led. One such example is the Takapuna Beach Cup, held annually in Auckand, New Zealand. The event was started in 2007 by a local community group, but has now grown significantly, attracting 289 teams in 2020 (Takapuna Beach Cup History, 2021).

At this level, there may be some level of tension between community events and the commercialisation of those events for event tourism reasons. However, they may have a role to play in temporally spreading economic gains as part of full event portfolio.

The final type of event in Getz's typology are *local events*, which have the lowest demand from non-local event tourists. These events are small in size and most likely led by the local community. There will be minimal tourism gain from these events as they may only appeal to the local community. For this reason, these events will have little impact on a destination's temporal spread or concentration of tourism.

Within each of these four categories are a variety of different events sectors. They could be mainly characterised under three areas: sports, culture and business. Ziakas and Costa (2011) suggest that sport and culture events have a symbiotic relationship.

Sports events

Sports events can exist at every level of the event typology from local events like parkrun right through to international mega-events such as the Olympic Games. From a temporal perspective, sports events are often cyclical. For example, parkrun is weekly (every Saturday). Major sports championships are usually annual, while mega-events like the Olympic Games are only held every four years. Many larger sports events that fall into the hallmark or mega-event categories move from place to place, often through a competitive bidding process. Destination managers seeking to fill gaps in their calendars may choose to enter this competitive bidding process to ensure a consistent event tourist provision. Higham (1999) argues that smaller sports events may have an overall longer-term effect with less reliance on the risks associated with building new infrastructure.

Cultural events

Cultural events are typically arts or entertainment-based and include a wide range of different industries. Some examples of cultural events are:

- Food and drink festivals (e.g. Oktoberfest)
- Arts festivals (e.g. Burning Man Festival)
- Music festivals (e.g. Glastonbury Festival, Coachella)
- Literature festivals (e.g. Hay Festival of Literature)
- Cultural celebrations (e.g. Chinese New Year)
- Religious celebrations (e.g. Diwali)

Cultural events are less likely to move from place to place, although there are notable exceptions to this, such as the European City of Culture. As with sports events, cultural events often occur on a cyclical basis (most commonly every year). Cultural events often have historical or religious connections to their place, which occur regularly, such as Chinese New Year or Diwali. Often destinations use these culturally time-defined celebrations to fill an event calendar, for example, Christmas markets during the month of December.

Business events

Also known as Meetings, Incentives, Conferencing and Exhibitions (MICE), in 2018, business and professional travel accounted for 12% of all worldwide trips (UNWTO, 2020). According to Getz (2013) MICE is the most developed component of event tourism and has temporal cycles. Most conferences and exhibitions happen on a yearly basis and some can also be won through private bidding processes. This sector is primarily driven by the need to fill venues and spaces throughout the year. One of the downsides to business travel is that the individual travellers are often constrained by time to explore a destination beyond the limited time boundaries of the event.

Temporal issues of event tourism

While a destination may itself be subject to pervasive temporal issues such as seasonality, there are a wide variety of other temporal issues at play with any event tourism strategy. Not all events will contribute to the tourism economy of a destination. Some local events lack appeal to non-local visitors, so contribute very little to overcoming temporal challenges in the visitor economy. Several other temporal influences on any event tourism programme need to be considered before implementing any strategy.

Supply-side temporality

The concept of supply-side temporality considers any event tourism strategy from the perspective of time and space available within a destination. Destination managers seek to fill gaps within their programme to suit venue availability, rather than consider the needs and demands of potential attendees. This supply-side approach requires that venue spaces are filled by events in order to cover overheads and running costs. This places additional pressure on the development or bidding of different events. Getz (2008) suggests that a supply-side approach remains the prevalent philosophy in most destination event tourism strategies. This can lead to copying other destinations' approaches and a lack of diversity in the event offering. For example, many destinations have a food festival, an arts weekend, a family music event and so on, which may compete head-on with those in other destinations.

Demand-side temporality

The concept of demand-side temporality considers the attendees from the start. From this perspective, any event tourism strategy considers the potential event attendees' needs first over any available time and space. From a temporal perspective, this is a challenge given that destinations have an overview of the time and space, and the needs or demands from attendees may not match the availability the destination. For example during school vacation period, the need for more family-based events or festivals is high, but the destination may have limited time and space to accommodate these. Getz (2013) advocates that a demand-side approach would offer the most significant competitive advantage, but that typically destinations follow the supply-side approach to creating events.

Temporal attractiveness

For many people, temporality is part of the attractiveness of events, in other words the ability to experience a single moment in time that will never be repeated. This has led to a new popular culture concept known as the "fear of missing out" or FoMO, which has proven to be a highly useful marketing tactic for selling events (Hodkinson, 2019).

Destination events are by their nature limited in duration and therefore, perishable. Once an event or festival has finished, it cannot be reused or experienced again. The time and resources needed by an event is a one-time use. This is particularly true of mega-events like the Olympics, or the European City of Culture where intense bidding occurs at great cost. The return required from event tourism is significant and so carries high economic and political risk. Perhaps the attractiveness of events, and indeed the greater value, comes from repeating events where a sustainable relationship between attendees and events can be developed over time. This relationship can be built on predictability or reinvention each time the event is delivered.

Political time frames

Traditional destination marketing organisations do not always have direct control over the services and products produced. They are attempting to tie together a wide range of services into one marketable product which can often be misaligned (Kavaratzis, 2004). One of the challenges destinations face is setting the direction of any event tourism strategy, which local political interests can highly influence. Political interests can change in step with the term of office for elected officials. When new officials are elected, priorities change. Unfortunately, this can often clash with the operational delivery of a long-term event tourism strategy. Newly elected politicians may be keen to attract high-profile events as a means for improved support, even if it goes against the long-term strategy. This political cycle can be a considerable challenge for destination managers, placing them in a difficult position with ever-changing demands.

Also, there are competing voices for both economic and social fulfilment. A destination may have culturally significant festivals and events each year. It is not always advisable to attempt to leverage these for tourism; thus, it may introduce significant tensions or disenfranchisement with the local community (Fredline & Faulkner, 2000; Roche, 1994).

Temporal strategies in event tourism

Destinations need to consider how their strategy towards event tourism is impacted by the effects and challenges of temporality. There is a growing body of academic literature around event tourism strategy and how it can overcome those challenges. Despite this growing knowledge, Norman and Nyarko (2021) found that less than 50% of towns and cities in the UK have an event strategy in place for their destination. The final section of this chapter will explore some of these ideas in relation to temporality.

Event portfolios

Event portfolios can be described as the strategic programming of events throughout a year, although the word 'portfolio' suggests assets taken from a financial sense. The benefits of such an approach include mitigating the negative impacts of seasonality, boosting economic benefits through leveraging and building synergies with social outcomes (Ziakas & Costa, 2011). Event portfolios are an emerging concept that an increasing number of destinations are seeking to exploit (Ziakas, 2019).

Events can become assets of value to a destination and sources of repeatable income. A destination plans and uses a portfolio to manage those assets to extract maximum benefit from events on a consistent basis, negating any temporal concentration effect caused by seasonality or cycles. Destination planners take a strategic viewpoint across a year to understand where there may be gaps in any event calendar and ensure they can leverage each event to full economic and social effect. Leveraging is an important component of event portfolios as it ensures that the local economy is fully aware of how to maximise its own rewards from each event.

In keeping with the supply-side challenges of temporality is the programming of events around the year to maintain a constant and sustainable event tourism portfolio. As previously mentioned, repeating events (annually or at other frequencies) builds up loyalty and social capital (Higham, 1999) that would make them more sustainable by reducing the temporal economic risk of single, one-off large events. Portfolios of events come into being either naturally or by design. In a natural organic way, events are created without any real consideration being given to the holistic picture for event tourism in the destination. An example would be how a local entrepreneur might start their own local annual music festival. In every sense this event is organic, with little consideration given to the wider tourism objectives of the host destination. Over time, the festival may attract tourism to the destination, but does not form part of the marketing 'offer' for the destination.

Bottom-up and top-down strategies

As previously highlighted, there are challenges in developing event tourism portfolios to overcome temporality issues. As Getz (2008) suggests, the most common approach to event creation in destinations is supply-side, or top-down. Top-down event tourism is a more managerial approach to developing an event tourism strategy. A policy document is created that describes how the destination plans to use events as a tourism tool. If a destination is aware of temporality issues such as seasonal demand peaks and troughs, social and environmental impacts of concentration and so on, it could use this approach to spread the supply of events across the year.

The advantages of using a top-down approach means that the policy-making body (be it local government, the DMO or similar) has complete control over the creative curation and distribution of events across the event tourism calendar. They can instigate events that suit their destination brand (Anholt, 2007) and the issues that the destinations faces. A top-down approach will require resources to deliver the event programme, and this will largely take the form of economic investment in resources such as staffing and physical infrastructure to support the events. They may also seek external economic investment through local businesses or sponsorship to support these events.

The downside to this approach is that authenticity may be lost. Authenticity can be a large part of the reason that people are attracted via tourism (Cohen, 1988). Artificial events that are created purely for the purpose of filling a space in a calendar are likely to be viewed as less authentic and so are less attractive to event tourists. There is also the risk that events will not meet attendee demands and will fail. Top-down strategies are not considered as responsive to changes in demand as they tend to have a lifespan of several years and are created and managed by local government organisations that often contain various levels of bureaucracy.

On the other hand, bottom-up event tourism is typically where events are created by locals, be they socially or economically-driven and on an individual or community basis. The events represent the people or place and so may be considered more authentic and attractive to would-be visitors. By taking this bottom-up approach, they also create a point of difference to stand apart from rival destinations (Andsager & Drzewiecka, 2002). Opportunities may arise to take advantage of temporally induced impacts, which requires swift intervention, something a top-down approach is less likely to be able to do.

One of the key challenges many destinations face is trying to tie together various products and services like events into one marketable brand. Edinburgh is an excellent example of how this can be done with its Festival City brand. Its event tourism portfolio is shaped around only festivals and does not attempt to focus its place marketing with business or sports events, although these still do happen.

Case study: Tramlines in Sheffield

Sheffield is a post-industrial city located in the north of England. It had a highly successful steel industry during the industrial revolution, but this has since declined (made famous by the film *'The Full Monty'*) leaving the city with significant economic and social challenges. In more recent times, the city has tried to re-invent itself as an events destination, primarily by hosting sports events (Bramwell, 1997; Roche, 1994). These events started with the World Student Games (WSG) in 1991 where Sheffield City Council funded the construction of several major venues including Don Valley Stadium, Ponds Forge and Sheffield Arena. However, it proved challenging to maintain a successful portfolio of events through the year to support all of these venues, given that the WSG made a financial loss and left the city with a reported debt of £659m (BBC, 2011).

The Head of Major Events and Markets for Sheffield City Council, Richard Eyre, describes the challenges associated with events in the city (R.Eyre, personal communication, May 18, 2018). The city does not attract day visitors; no coaches arrive bringing visitors, and retail is in decline. The city has tried to move away from solely hosting sports events due to the inherent risk of bidding for events. Instead, the city focuses on developing events that can repeat each year as a path towards greater stability. Richard suggests that the council plays the role of a facilitator supporting grassroots community groups or entrepreneurs in the private sectors to develop these homegrown 'temporally secure' events within the city.

Tramlines is an annual music festival held in Sheffield. The first festival was held in 2009. The name is a nod to the light-rail tram system around the city centre. It was a collaboration between local venue owners, promoters and the Sheffield local government. As a major university city, home to over 60,000 students, Sheffield often became noticeably empty during the student summer vacation and so this group looked at ways to boost footfall into the city centre during this low season summer period.

The festival occurs in late July when there is less likely to be students using local bars and music venues. This time is also when family holidays start. As Sheffield does not have a typical tourism economy with a critical mass of established attractions, this exodus of students and local residents has resulted in a negative footfall to the city during summer months.

Figure 13.2: Tramlines Festival, 2018. Photo: Fanatic_GilesSmith_Tramlines

Since 2009, Tramlines has become a cultural Hallmark event for Sheffield, boosting the local economy during a traditionally quieter period. Despite this success story, Richard Eyre suggests spotting future star events is problematic as not all events have tourism potential, nor can they dovetail into their event calendar.

Tramlines Music Festival: https://tramlines.org.uk/

Conclusion

This chapter has demonstrated that the relationship between event tourism and temporality occurs on many levels. There are both supply and demand issues that often mean strategies by destinations must be carefully considered to ensure a successful outcome. Event tourism offers destination managers an opportunity to spread visitor demand throughout the year through innovative and creative events, but often the strategy is driven by the need to repeatedly fill venues and spaces. Attractiveness and political forces come and go with time, but a long term strategy to embrace the development of a sustainable event portfolio offers the best opportunity to balance a destination's temporal challenges.

Self-reflection questions

1. What features of events cause them to be 'temporal'?
2. Which type of events are most suited to becoming tourist attractions in their own right?
3. Why do tourism organisations like DMOs often use a top-down approach to event planning?
4. How might temporal issues in a destination be overcome by using a bottom-up event tourism strategy?

References

Andsager, J. L., & Drzewiecka, J. A. (2002). Desirability of differences in destinations. *Annals of Tourism Research,* **29**(2), 401-421. doi:10.1016/S0160-7383(01)00064-0

Anholt, S. (2007). *Competitive identity: The new brand management for nations, cities and regions.* UK: Palgrave Macmillan.

BBC. (2011). Sheffield's world student games £658m debt 'disaster'. https://www.bbc.co.uk/news/uk-england-south-yorkshire-14134973

Bramwell, B. (1997). Strategic planning before and after a mega-event. *Tourism Management,* **18**(3), 167-176. doi:10.1016/S0261-5177(96)00118-5

Bramwell, B., & Rawding, L. (1994). Tourism marketing organisations in industrial cities. *Tourism Management (1982),* **15**(6), 425-434. doi:10.1016/0261-5177(94)90063-9

Cohen, E. (1988). Authenticity and commoditisation in tourism. *Annals of Tourism Research,* **15**(3), 371-386. doi:10.1016/0160-7383(88)90028-X

Connell, J. (2015). *Visitor attractions and events: Responding to seasonality.* New York: Elsevier Science.

Connell, J., Stephen, J. P., & Meyer, D. (2015). Visitor attractions and events: Responding to seasonality. *Tourism Management,* **46,** 283-298. doi:10.1016/j.tourman.2014.06.013

Crouch, G. (2010). Destination competitiveness: An analysis of determinant attributes. *Journal of Travel Research,* **50**(1) doi:10.1177/0047287510362776

Fredline, E., & Faulkner, B. (2000). *Host community reactions.* Oxford: Pergamon. doi:10.1016/S0160-7383(99)00103-6

Getz, D. (1997). *Event Management & Event Tourism.* New York: Cognizant Communication Corporation.

Getz, D. (2008). Event tourism: Definition, evolution, and research. *Tourism Management,* **29**(3), 403-428. doi:10.1016/j.tourman.2007.07.017

Getz, D. (2013). *Event Tourism : Concepts, international case studies, and research.* Putnam Valley, NY: Cognizant Communication Corporation.

Getz, D. (2016). *Progress and prospects for Event Tourism Research.* New York : Elsevier Science.

Higham, J. (1999). Commentary - sport as an avenue of tourism development: An analysis of the positive and negative impacts of sport tourism. *Current Issues in Tourism,* **2**(1), 82-90. doi:10.1080/13683509908667845

Hodkinson, C. (2019). 'Fear of missing out' (FOMO) marketing appeals: A conceptual model. *Journal of Marketing Communications,* **25**(1), 65-88. doi:10.1080/13527266.2016.1234504

Kavaratzis, M. (2004). From city marketing to city branding: Towards a theoretical framework for developing city brands. *Place Branding,* **1**(1), 58-73. doi:10.1057/palgrave.pb.5990005

Lai, K. (2016). Influence of event image on destination image: The case of the 2008 Beijing Olympic Games. *Journal of Destination Marketing & Management,* **7**(1), 153-163. doi:10.1016/j.jdmm.2016.09.007

Oh, M-J. & Lee, T. (2012). How local festivals affect the destination choice of tourists. *Event Management,* **16**(1), 1-9. doi:10.3727/152599512X13264729827479

Norman, M., & Nyarko, N. (2021). Networked economic value creation in event tourism: An exploratory study of towns and smaller cities in the UK. *Event Management,* **25**(1), 69-83. doi:10.3727/152599520X15894679115493

Pacione, M. (2012). The role of events in urban regeneration. In S.J. Page & J. Connell (Eds), *The Routledge Handbook of Events*. Abingdon: Routledge.

Pugh, C., & Wood, E. H. (2004). The strategic use of events within local government: A study of London borough councils. *Event Management*, **9**(1-2), 61-71. doi:10.3727/1525995042781093

Richards, G. (2017). Emerging models of the eventful city. *Event Management*, **21**(5), 533-543. doi:10.3727/152599517X15053272359004

Richards, G. & Palmer, R. (2010). *Eventful Cities: Cultural Management and Urban Revitalisation*. Oxford: Elsevier.

Roche, M. (1994). Mega-events and urban policy. *Annals of Tourism Research*, **21**(1), 1-19. doi:10.1016/0160-7383(94)90002-7

Takapuna Beach Cup History. (2021). www.takapunabeachcup.com/history

UNWTO. (2020). Global and regional tourism performance. https://www.unwto.org/unwto-tourism-dashboard

Ziakas, V. (2019). Issues, patterns and strategies in the development of event portfolios: Configuring models, design and policy. *Journal of Policy Research in Tourism, Leisure and Events*, **11**(1), 121-158. doi:10.1080/19407963.2018.1471481

Ziakas, V., & Costa, C. A. (2011). The use of an event portfolio in regional community and tourism development: Creating synergy between sport and cultural events. *Journal of Sport & Tourism*, **16**(2), 149-175. doi:10.1080/14775085.2011.568091

14 Marketing the Seasons

Richard Tresidder and Emmie Deakin

Learning outcomes

After reading this chapter, you will be able to :

1. Understand how the season can be utilised to extend both the value and volume of tourism.
2. Explore the concept of temporal marketing
3. Investigate the semiotic seasonal language of tourism
4. Identify the role of semiotics in tourism marketing

Introduction

This chapter explores the use of seasons in defining the marketing offer for destinations. Rather than examining the traditional idea of seasonality within the tourism industry that is formed around both weather conditions (skiing in winter, beaches in summer etc.) and calendar events (school holidays, Spring break or religious holidays etc.), this chapter explores how the cycle of seasons are linked to specific touristic experiences that help to reduce the impact of seasonality on businesses by creating a continuous sequence of events throughout the year.

Tourism seasonality is generally seen as a negative element of tourism with many workers only being employed for short periods of time (Duro & Turrión-Prats, 2019). Within this context Butler (1998) refers to the problem of seasonality as a 'temporal imbalance' in the phenomenon of tourism.

Although the issue of seasonality is traditionally seen as problematic for many organisations and destinations, many have effectively utilised the seasons and characteristics of these to creating an ongoing, effective and creative marketing campaign. As will be seen within this chapter, one of these organisations is the National Trust in the UK and specifically, Calke Abbey in Derbyshire will provide a case study of how the seasons may be used as the central theme of a marketing strategy.

Case study: The National Trust at Calke Abbey

The National Trust for Places of Historic Interest and Natural Beauty (known as 'The National Trust'), is an independent registered charity reliant on income from membership fees, donations/legacies and revenue from their commercial operations, such as tea rooms and shops. The National Trust was established by social reformers in 1895, to act as guardian for the permanent preservation of places and areas which are historically significant or areas of natural beauty, for the benefit of the nation and across England, Wales and Northern Ireland (Dickinson et al., 2004; Lithgow & Timbrell, 2014). The National Trust is now one of Europe's largest conservation charities and is custodian of more than 500 historic houses, parks, gardens and ancient monuments; 780 miles of coastline; 250,000 hectares of land and approximately a million works of art (National Trust, 2021).

This chapter will focus on Calke Abbey, a grade 1 listed Country House in the care of The National Trust, located in Ticknell, Derbyshire in the East Midlands region of the United Kingdom. Calke Abbey was once the ancestral home of the Harpur Crewe family, who were eclectic and avid collectors, but the estate is atypical of country houses within the care of the National Trust. Referred to on the entrance sign as 'The unstately home', it is as if time has stood still to reveal peeling wallpaper and paintwork and rooms stuffed full of collections. The house and stables are little restored. This was a deliberate move by the National Trust who decided to consolidate and preserve the house, stables and collection as found, rather than restore to its former glory. The house and gardens illustrate the story of the decline of the country house estates. In the early-mid 20th century, the survival of many of them were threatened due to inheritance taxes, staff shortages and a change in the social fabric of the country during that period and the way people lived and worked.

Whilst the house closes for winter, the key element for the purpose of this chapter is the 600 acres of parkland within which the house and stables sit. Within the parkland there are woodlands and ancient trees, a nature reserve, long horn cattle, rare breed sheep and 67 acres of deer park which is home to red and fallow deer. Social media (Facebook, Twitter and Instagram) promote the seasons and temporal activities all the year around, whether this be an image of a sunset over a lake representing the longer days as winter turns into spring, but also the tranquillity and peace and quiet that many of us crave; the first snowdrops poking their heads through the snow; the impressive and annual display of bluebells within the woods; the Gourd display in the gardens, the striking autumnal colours of the leaves and the rutting deer in the autumn; the house being illuminated for the festive season.

Image 1: Calke at Christmas

Image 2: Bluebells in the wood at Calke

Image 3: Autumn at Calke. (Photos by the authors)

Temporal marketing

Tourism marketing has always been driven by the climatic, cultural and seasonal characteristics of a place, with the sun, snow and annual festivals being utilised as the principal component of a marketing strategy. For many destinations this has led to 'on and off' seasons (Gkarane & Vassiliadis, 2020), with hotels shops and restaurants being closed for large parts of the year. For example, the British seaside holiday season is limited to approximately sixteen weeks a year. What this means in practice is that tourism businesses must generate all their income during this period (Ferranteet al., 2018) and in the off season there is little income, elevated levels of unemployment and, in some cases, poverty and deprivation (Martin Martin et al., 2020). This limited season is also the case for many other destinations and as such there has been an ongoing search for new and innovative marketing strategies and activities to extend the visitor season. Successful examples of this are the rise of Christmas markets, food festivals, cultural events and the conference market, all of which take place out of high season and make use of the availability of hotel rooms and exhibition spaces.

The purpose of such product development and marketing activities is to increase the volume and value spread of the tourism product, thus creating a more financially sustainable product that is not so reliant on peak periods. Apart from increasing tourism revenue, the spreading of the season also reduces some of the risk factors such as harsh weather, lack of snow or labour strikes impacting upon tourism businesses (Chen et al., 2019). The case study offered in this chapter demonstrates how an effective marketing campaign and strategy can effectively increase the length of the tourism season by creating multiple mini events that represent the changing seasons, which utilise a form of temporal positioning that creates a more holistic and sustainable product.

If we explore the way in which the tourism offer is constructed at Calke Abbey, we see that the house forms a vital role of framing the touristic experience. It provides the identity and history of the site; it forms a focal point around which visitors congregate and is central to contextualising the marketing for the site. All the practical tourism infrastructural elements such as toilets, café, shop and carparking are located together and adjacent to the house. So, for most visitors even when the house is closed for winter, their experience of the parkland and gardens starts at this central point. Thus, the value of the house remains an important commodity as a contextualisation point for the visit as it forms a background to the seasons. For example during the Christmas holiday period, the exterior of the house is lit up (see Image 1), you are invited to walk through the gardens to follow the trail of twinkling lights through to the church and into the gardens. Whilst the house may be mothballed over winter, access to the park and grounds ensures that there is a steady income generated through admission fees, specific events and income from the National Trust shop and café. As the National Trust relies on the recruitment of paying members, the extension of the season also augments the value of being a member and not just a day visitor. As a result, many of the jobs in servicing tourists are full time rather than just seasonal. The utilisation of the season to extend the value and spread of tourism creates a calendar of events (whether natural or organised) that enables the effective year-round tourism activity at the site.

Although all destinations can engage in seasonal marketing, properties such as Calke Abbey possess a competitive advantage over other sites in terms of the amount of cultural and natural resources they possess. Apart from the house, the church, deer house, riding school building and stables all provide resources to stage the Calke estate as a working estate, with farm animals, deer and an abundance of indigenous flora and fauna (see images 2 and 3). This provides them with a wide range of opportunities to use these as the centre of any marketing campaigns, each creating a mini-event that forms a rich catalogue of touristic experiences for the local urban population. As can be seen in the images of Calke, mini-events include bluebell walks, the deer rutting season and the Autumnal leaves, the arrival of spring lambs. In addition, the availability of covered areas such as the riding school, also offer a space for seasonally orientated food festivals, Christmas markets, concerts and historic reconstructions. Each National Trust property has a differing set of resources that enable differentiated touristic experiences while adopting similar seasonal marketing strategies.

What this case study demonstrates is that the use of seasonal marketing can be utilised to extend the period in which tourists are motivated to visit a destination. Although this case study concentrates on a single site, the lessons from it can be implemented at both regional and the national level. Every destination has a particular set of natural, cultural and physical resources that may be utilised to create unique experiences for tourists to explore (Urry & Larsson, 2011). This creative use of the seasons moves away from the concept of 'off' and 'on' seasons but goes further to develop a portfolio of opportunities for visitors to engage in, both targeted interest group activities and to make effective use of the growing mini break market. The seasonal marketing of a destination has witnessed the growth of targeted marketing campaigns both in brochures and particularly effectively in the realm of social media, where it is possible to quickly embed the emergence of flowers or the sighting of animals in postings. Consequently, these utilise a set of signs and images that we, as the consumer, can identify with and understand.

Temporal semiotics

Tourism marketing utilises a semiotic language of tourism that communicates both experiences and expectations to the potential consumer (Tresidder, 2014). Semiotics can very simply be understood as the study of signs and images, how they generate meaning and how we as consumers interpret and find meaning through the reading of them. The signs and images used within tourism marketing draw from a set of words, pictures and phrases that have a connotation particular to the tourism industry. For example, just as a picture of the Eiffel Tower is instantly recognisable, the tower is used by tourism marketers to represent Paris or France and even love and romance (Urry & Larsen, 2011), the image of bluebells or lambs come to represent spring, rebirth and the end of the bleakness of Winter.

The experience of consuming tourism is not a tangible activity. Unlike buying a car we cannot test drive, touch or even smell a holiday or event before we purchase it. As such it is important that we communicate the essence of the destination through marketing activities and the expected experience to potential tourists or visitors (Hirst & Tresidder, 2016). This is achieved by utilising a set of signs and images that we recognise or that create a curious desire within us. The various forms of marketing communications such as brochures, television adverts, web pages or social media postings all act as the 'sign vehicle': that is they transport messages to us the consumer. Each of these messages are rich in meaning. These communications are not just arbitrary, but form part of a complex set of signs and images that consumers understand and they mean something to us all (Tresidder, 2015). In short, we can all relate to the various seasonal symbols presented in marketing materials. For example, the Austrian Tourist Board presents a picture of the various markets as a significant element of the Austrian Christmas tradition (Bigné & Decrop, 2019). They utilise a set of images that use historic buildings as backdrops, with lit Christmas trees, visitors drinking gluhwein and lots of warmly clad families (see Bausch & Unseld, 2018). It is a picture book image of Christmas, reinforced by the text which asserts that.

Christmas Markets in Vienna

Step into the joy of Christmas with the beautiful streets of Vienna as your backdrop. Punch and chestnut stands draw customers with seasonal treats and stallholders get shoppers in the Christmas spirit with handcrafted goods.
(www.austria.info/en/things-to-do/skiing-and-winter/christmas-markets/vienna)

What this statement achieves is to add a further authenticity to the experience. In fact, the promotional material effectively gathers all the elements associated with a traditional Christmas experience. This concept of authenticity is even more explicit in the promotional material for the region of Carinthia, stating:

The time before Christmas sprinkles a distinct magic across Carinthia. Atmospheric Christmas Markets invite with advent concerts, authentic handicrafts and traditional customs. Enjoy the thoughtful side of the pre-Christmas period.
(www.austria.info/en/things-to-do/skiing-and-winter/christmas-markets/klagenfurt-carinthia)

The relationship between the images and the text invites the visitor into an authentic and traditional representation of Christmas that is understood by many tourists no matter their ethnicity, religious beliefs or age (Bausch & Unseld, 2018). It is a notion of Christmas that has been embedded in our psyche and is played out in many films, television programmes and books.

It can be argued that the tourism industry utilises different sets of signs and images to not only represent the seasons, but also to represent distinct types of experiences (Urry & Larsen, 2011). Tourism marketing is in fact underpinned by a semiotic language of tourism, and which we will return to later in this chapter. For example, Austria's website splits outdoor activities into Summer and Winter, with images of snowy streetscapes and scenescapes, families in the snow, ice skating and even horse drawn sled rides, whilst activities such as hiking, and biking are always set in sunny summer days. What we witness is that the available touristic resources become associated with different activities, rather than just thinking about sites as a skiing destination (Bausch & Unseld, 2018). The value and spread of the resources are therefore commodified to create an ongoing seasonal tourism offer. The main website is planned in advance and has to a degree replaced the brochure, thus the images and the strategy have to be fixed ahead of time. However social media provides organisations with the ability to effectively represent the vagaries of the natural environment. For example, if lambing is early or there is an unusually high number of bluebells, all of these become micro-events that can be marketed and communicated quickly and effectively through social media. As can be seen from the case study, the National Trust has become very adept in harnessing nature and the seasons as a set of events that can be adopted to attract tourists, ensure members feel an affiliation to their sites and removes pressure from an over reliance on just the historic house alone. However, in order for this to work they need to draw from an extensive and embedded set of signs and images, which form the semiotic language of tourism. It is to this we now turn.

The seasonal semiotic language of tourism

As the interpretation process involves a certain degree of emotional involvement, we need to understand how the semiotics of tourism and events builds this relationship with the consumer. There are a number of conventions that underpin definitions of contemporary tourism and events and include certain themes and signs that construct the experience (Hirst & Tresidder, 2016). The seasonal marketing of tourism can be broken down into three types: macro seasonal events, micro seasonal events and annual cultural events. Each of these draw upon a different semiotic strategy within the various forms of marketing communication. Furthermore, each of these strategies employ different types of images ranging from the generic to the temporally and geographically specific. The first element of the language of tourism and events marketing involves the semiotic construction of a time and place in which the experience is located. The images utilised in all the categories offer the potential tourist or visitor entry into a time and space that is removed from everyday lived experience or can be perceived as 'extraordinary' (Urry & Larsen, 2011). This is demonstrated in the use of phraseology in marketing that focuses on the use of emotive words and images that represent the seasonal, temporal and significance of the event.

Table 14.1 : Seasonal semiotic language and representation in tourism

Semiotic language	Semiotic representation
Macro seasonal events Summer season Winter season Spring season Autumn season	Macro events are the large season events which encompass large periods of time. For example, the idea of summer sun in the Mediterranean may be between April and September. Images will generally include beach scenes/countryside, sunshine and groups of families. These utilise very generic pictures that represent the entire season. The purpose is to appeal to as wide an audience as possible. Autumn in Vermont, Spring in Brittany, Winter Wonderland in Switzerland etc. All these examples form part of macro seasonal events
Micro seasonal events Lambing Bluebells Turtles laying eggs Deer rutting Apple harvest	These are specific temporal events; these are generally communicated through social media and utilise pictures of the event as they are unfolding. The images used are generally specific to the event but utilise certain conventions. For example, pictures of bluebells are generally shown within a deserted forest context. This contextualises and reinforces the naturalistic and bucolic significance of the scene and the event. Such vignettes enable us to link and associate with the seasons. The marketing enables us to identify the availability of tourism resources to physically interact with these micro season events.
Annual cultural events Easter Eide Christmas Independence Day Sukkot Chinese New Year	These events mark the seasons and form traditional temporal markers as to the passing of the year. Images used within promotion focus of the characteristics of the event, for example the Christmas Markets in Vienna identified above or Easter Eggs and Bunnies for Easter. This approach draws on generic and culturally embedded definitions and representations of the event. This differs from macro events as they are temporally and culturally specific.

The seasonal semiotic language which is used to construct and inform visitors is rich in naturalistic, temporal and cultural significance. It not only represents the event itself, but it is also rich in meaning, it represents something. For example, images of bluebells or lambs represent rebirth and the emergence from winter. It marks the change in seasons and, through tourism, the opportunity to tangibly consume it. The semiotic marketing of season forms an important role in allowing destination to form a seasonal marketing strategy that makes use of their cultural and geographical capital in order to increase the value and spread of their tourism product.

Summary

This chapter has identified that the development of tourism need not be constrained by the traditional ideas of 'on' and 'off' seasons, but that each country or region possesses a set of cultural, geographical and seasonal resources that can form the effective foundations of a marketing campaign. The Calke Abbey case study demonstrates that although the main attraction is closed for extended periods, the surrounding resources can create experiences for visitors throughout the year. The outcome of this is that jobs can be seen a permanent, members have resources they can access all year and visitors can be attracted no matter what the season, thus providing a sustainable financial income. Calke Abbey has been creative in utilising a set of macro, micro and annual events to develop a calendar of activities that appeal to all demographic groups. This development has been supported by a meaning semiotic language that effectively communicates the experiences to the potential visitor.

Self-reflection questions

1. Reflect on how your region could or has developed a seasonal marketing strategy.
2. What semiotic strategy would you use to market your chosen destination?
3. Undertake a cultural and geographical audit of the resources in your locality and identify how these could be packaged as a year-round tourism product

References

Bausch, T., & Unseld, C. (2018). Winter tourism in Germany is much more than skiing! Consumer motives and implications to Alpine destination marketing. *Journal of Vacation Marketing*, 24(**3**), 203-217.

Bigné, E., & Decrop, A. (2019). Paradoxes of postmodern tourists and innovation in tourism marketing, in E. Fayos-Solà & C. Cooper (eds.) *The Future of Tourism*, Springer, Cham. pp. 131-154.

Butler, R. (1998), Seasonality in tourism: Issues and implications, *The Tourist Review*, 53(3), 18-24.

Chen, J. L., Li, G., Wu, D. C., & Shen, S. (2019). Forecasting seasonal tourism demand using a multiseries structural time series method. *Journal of Travel Research*, 58(1), 92-103.

Dickinson, J., Calver, S., Watters, K. & Wilkes, K. (2004) Journeys to heritage attractions in the UK: A case study of National Trust property visitors in the south west. *Journal of Transport Geography*, 12(2), pp 103-113.

Duro, J. A., & Turrión-Prats, J. (2019). Tourism seasonality worldwide. *Tourism Management Perspectives*, 31, 38-53.

Ferrante, M., Magno, G. L. L., & De Cantis, S. (2018). Measuring tourism seasonality across European countries. *Tourism Management*, 68, 220-235.

Gkarane, S., & Vassiliadis, C. (2020). Selective key studies in tourism seasonality: a literature review. *Strategic Innovative Marketing and Tourism*, 247-256.

Hirst, C. & Tresidder, R. (2016) *Marketing Tourism, Events and Food: A Customer-based approach* (2nd ed.). Goodfellow Publishers, Oxford.

Lithgow, K. & Timbrell, H. (2014) How better volunteering can improve conservation: Why we need to stop wondering whether volunteering in conservation is a good thing and just get better at doing it well. *Journal of the Institute of Conservation*, 37 (1), 3-14.

Martín Martín, J. M., Salinas Fernandez, J. A., Rodriguez Martin, J. A., & Ostos Rey, M. D. S. (2020). Analysis of tourism seasonality as a factor limiting the sustainable development of rural areas. *Journal of Hospitality & Tourism Research*, 44(**1**), 45-75.

National Trust (2021) National Trust Annual report 2020/21. https://nt.global.ssl.fastly.net/binaries/content/assets/website/national/pdf/nationaltrustannualreport2020_21.pdf

Tresidder, R. (2014). The semiotics of tourism marketing. In S. McCabe (ed.) *The Routledge Handbook of Tourism Marketing*, Routledge, pp. 116-128.

Tresidder, R. (2015). Experiences marketing: A cultural philosophy for contemporary hospitality marketing studies. *Journal of Hospitality Marketing & Management*, 24(**7**), 708-726.

Urry, J., & Larsen, J. (2011). *The Tourist Gaze 3.0*. Sage, London.

15 Planning for Seasons: Value Chain Management and Digitization

Jana Heimel

Learning outcomes

The chapter will enhance your appreciation of:

1. The relevance of capacity planning for seasonal planning
2. Solutions for supply chain contracting
3. How to build up reputational capital
4. Digitalization for seasonal planning

The purpose of the chapter is to provoke students to think about the wider implications of temporal variation as challenges for operational planning for tourism businesses. In order to achieve this, this chapter will provide holistic perspectives and case studies from a range of players within the tourism related industries: e.g. tour operators, cruise line operators, transport operators, visitor attractions.

The chapter demonstrates how value chain management supports planning for seasonal variations and how important the role of digitalization is.

Supply chain management for generating added value

Value chain management

A firm's value (added) chain maps the entire transformation process of input into output across all its activities. It represents the stages of production as an orderly sequence of a firm's activities. These activities, creating value by consuming resources (input variables), are linked in processes transforming the input into valuable goods or services (output variables) (Porter, 1980). The analysis of the value chain gives a starting point for creating strategic competitive advantages. The difference between the costs of value creation activities and the customer benefit expressed in the market price forms the profit margin. Each value activity can be used as the basis for identifying and generating cost advantages or sources of differentiation.

Managing the value chain of tourism organizations belonging to the service sector significantly differs from the manufacturing industries. Specific service sector characteristics include intangibility, heterogeneity, customization, integrativity, interactivity, lack of transparency and relatively high overheads (Meffert, 2000; Meffert & Bruhn, 2000; Witt, 2003; Sethi, 2017). Whereas in the production industries value creation highly depends on machines and facilities, in the service sector human capital and know-how represent the main capital (Maleri, 1997; Maleri & Frietzsche, 2008; Pasban, 2016). However, there are tourism businesses whose services also rely on the provision of buildings or other facilities, such as hotels, airlines or cruise liners.

For this reason, there is a need for a special focus on capacity planning including (human) resource and working capital management as well as supply chain contracting and the generation of reputational capital whilst planning for seasonal variance.

Capacity planning and price differentiation

There are two main challenges in relation to human resource management and planning for temporal variations. On the one hand tourism businesses exposed to fluctuations between off-/low and busy seasons in some cases, even accompanied with laying-off staff during these times, are especially challenged to constantly deliver a high service quality when reopening after a break. For those it could be useful to offer team meet-ups, training and personal development in the off-peak season in order to stay in touch, to teach soft and hard skills and to transfer industry knowledge as well as expertise to further develop the organization and thus guarantee good service quality in the new season. On the other hand, tourism businesses with high sales volatility face a particular challenge when planning staff capacity. For example, a mountain hut operator in the Alps (from the German Alpine Association), only being opened in the summer time, will need to make use of fixed-term contracts with its staff and provide personal development opportunities in the valley to bridge off-/low-seasons.

Next to staff planning, the planning of capacities for other material resources such as in the case of Stuttgart by Bike (SbB – see below), rooms to store the bike fleet to rent out and immaterial resources (e.g. audio/VR technologies, booking software) implies another key success factor when it comes to planning for seasons. Unused resources represent opportunity costs. The perishability of such resources doesn't involve a 'real' disbursement, however it represents a loss of sales because of the opportunity that is lost. To reduce the probability of lost turnover, the calculation of special priced offers and bundled packages for off- or shoulder seasons can contribute to minimizing opportunity costs.

Case study: SbB – Stuttgart by Bike

SbB is a start-up business in Germany offering bicycle tours, rentals and courses. It is confronted with one low-busy season due to an intra-week cycle, and one off-season cycle when it comes to winter times. Whereas the weekends in the busy season are characterized by an above average demand for e-bikes with a risk of overbooking, the fleet mostly stays in stock during the weekdays thus generating a high working capital accompanied with high opportunity costs. To overcome these temporal variations and to balance revenues, SbB first had to identify these peaks by analysing past business data, second, estimate the approximate demand for future intra-seasonal cycles and third, calculate optimal prices to maximize revenue. Finally, SbB made use of bike rental fee variations during the day (e.g. special prices for students/retirees) to attract more customers and thus generate more revenue by reducing working capital.

Figure 15.1: Stuttgart by Bike opening hours (Offnungszeiten). Source: www.stuttgart-by-bike.com

Both staff and resource planning require an accurate capacity planning process with a continuous rolling demand forecasting supported by state-of-the art technological infrastructure.

Supply chain contracting

There are close links between the individual value activities and the value chains of suppliers, sales channels and customers, which can be used to create competitive advantages (Porter, 1985). Thus, supply chain contracting represents another key success factor within the seasonal operations framework.

To coordinate decisions along the entire supply chain, contracts are necessary. Such decisions include the choice of which inputs from which supplier as well as which outputs for which customers to use, transportation of both inputs and final products, inventory decisions at each stage and ultimate pricing. No single tourism firm controls all decisions along the supply chain, but many firms make decisions that must be coordinated through pricing and return policies, especially in for the case of perishable goods (Pasternack, 2008; Krishnan & Winter, 2012). The supply chain partner network should not rely on single sourcing (e.g. if a cruise liner buys food/materials from only one supplier) or single selling (e.g. if a business hotel rents out rooms only to a firm located close to the hotel) with regard to seasonal variations. In both cases each service provider risks revenue loss if its partner firm cancels the cooperation. For managing a supply chain successfully, flexible but sustainable contracts should be designed to be able to act and not just react, especially in case of occurrence of unforeseen situations. 'Best practice' firms are less reactive but rather act and handle potential challenges by implementing counter measures in advance to eliminate risks and prevent potential failures. General framework agreements can provide a solution as they define long-term prices and policies and thus grant security in uncertain times. However, they can also prevent or hinder flexibility. Partner contracts with high monthly fixed cost components and/or revenues being skimmed as a result of transfer or agency fees, can have a negative impact on profit, in particular with regard to seasonal variations.

Case study: AIDA cruise line operator

AIDA is a German cruise line operator. Maintenance of cruises generates high fixed costs, for example harbour berthing charges or maintenance costs for the cruise ship while in port. A clause with an option to defer contracts can help to survive low-season or off-season times (e.g. during the Covid-19 pandemic). AIDA decided upon cancelling all tours with the first wave without knowing how the situation would develop throughout the year (AIDA, 2020). The same accounts for hotels with a low occupancy rate or even no occupancy at all because of legally enforced shut-downs. Fixed costs for electricity and staff costs remain. Contracts with terms giving more flexibility on the scope of action can help to manage and plan for seasonal volatility appropriately.

Another very essential example for supply chain contracting occurs through the collaboration with distributors. They typically charge 10-20% for promoting tourism products on their platforms and channels (von Dörnberg et al., 2018). Care should be taken when entering contracts with intermediaries offering a sales platform along with one's own distribution channels. Since such intermediates typically ask for significant discounts from the selling price and in addition may charge high agency fees, the exact profit margin should be calculated and assured that it will be enough to cover fixed costs as well. Although such contracts can be useful for extending customer base, they bear the risk of unprofitability.

SbB signed contracts with a handful of vendor platforms such as Groupon, BajaBikes and Jochen Schweizer, to benefit from their marketing and sales expertise and wide reach. This enabled SbB to establish a profile in new and especially international markets. However, in busy seasons it caused overbooking and sales losses because of high commission fees and missing integration of the different sales channels.

These examples underpin the importance of accurate forecasting of demand as well as fully automated and integrated technological infrastructure.

Building up reputational capital

Tourism services are used by customers as processes. This procedural character leads to the need for contact between producer and buyer. Value activities therefore relate to personal, telephone, written or electronic contact options. Reputational capital is therefore becoming increasingly important for both the buyer and the supplier. In the area of marketing and sales, not only customer relationships need to be maintained and expanded, but also relationships with suppliers and investors.

For the service provider, the main result is to value activities in the form of branding, efficient design of the sales network and the expansion of the company image (i.e. an increase in awareness through advertising measures and media presence). These value activities serve to create reputation capital, which increases the loyalty potential of prospective customers and promotes lasting conviction with regard to the service provider's performance. Even after the actual production process, the service provider must maintain customer contact and customer satisfaction given the relevance of customer loyalty. Likewise, the provider must seek to avoid latent dissatisfaction and be active in the field of customer service.

Lots of tourism businesses nowadays make use of conversational marketing to be closer to their target group and to build up a relationship with their customers more quickly - regardless of the time of day or night.

> For example SbB uses chat boxes, chatbots and messenger services as well as its own social media channels for digital communication with customers. The user experience for their target group is increased by personalizing the digital chat conversations as much as possible, continuously testing the chat bots and adapting them to user needs. Moreover, private chat groups of messenger services (WhatsApp and Facebook Messenger for example) are

used to stay in touch with customers and quickly reply to their questions. By communicating through social networks, visitors can connect with experts leaving messages or comments.

Another option for promoting tourism products as well as services, and increasing customer loyalty is affiliate marketing (AM). The merchandiser supplies the affiliate with marketing material for publication and creates back-links. The affiliate charges a provision fee in form of pay per click/lead/sale or other models depending on the contract. In the tourism sector typically 15% up to 25% are charged as standard (Bormann, 2018). AM facilitates the establishment of new markets and thus generates higher sales. Besides, AM offers a good opportunity to collect reviews on other platforms beyond own business boundaries.

SbB for instance, promotes its tours and bikes on Google my Business and other intermediate platforms such as a local tourism agency, BajaBikes, Trip Advisor and meetups, to sell products globally and explicitly addresses international tourists. Special campaigns are launched on Facebook, Instagram and Google to reinforce the selling of products particularly in low/off seasons.

All the above aspects highlight the importance of an integrated and fully digitized information technology infrastructure and role of digitalization for planning for seasons.

Digitization

Digitization can be defined as a business transformation process determined by an increased use of information and communication technologies (Franken, 2016; Gartner, n.d.; Petry, 2016). Digital technologies (such as IoT, big data, mobile, VR devices and artificial intelligence, along with social developments (e.g. globalization, development of algorithms) drive this trend. It affects products and services as well as processes, working methods, business models and ultimately management approaches in their entire corporate management (Kreutzer et al., 2017: 1). Digitization has an impact on and is the foundation of all previous mentioned key factors for seasonal planning.

First, management decisions can be supported and implemented faster with an information and communication technology infrastructure based on data with one single point of truth combined with digitized analyzation methods and tools (Horváth et al., 2020). The use of large amounts of data from various sources with a high processing speed can generate enormous benefits.

The ability to collect large amounts of data (volume) from different sources and with different structures (variety) at high speed (velocity) to collect, process, and store data with good quality (veracity) and to evaluate the objective of an economic benefit (value) is referred to as 'big data'. The 5 Vs are also called the 'five characteristic features of big data' (Zikopoulos et al., 2013; Lippold, 2017). An integrated database with one single point of truth connecting internal and external sources of data, and assuring a standardized data structure are essential for generating big data analysis. Such analysis allows for a better and deeper understanding of business activities resulting in optimized management decisions. The connection of new data sources to management control systems and the further processing to relevant control parameters supports a more efficient and faster implementation of measures, especially when supplemented with the automation of data transmission that was previously carried out manually by software programs.

In order to be able to benefit from data (quantities), special data analysis procedures are used for the automatic recognition of patterns, meanings and interrelationships. This requires information technologies (IT) that support management in providing information. Business analytics aim to use analyses in a targeted manner to achieve set corporate goals. It combines the essential elements from data management, analysis methods and the presentation of results with the goal of continuous development and improvement (Davenport et al., 2010; Davenport & Harris, 2007). To apply these ideas to the SbB example, the application of business analytics empowers SbB to make better decisions, to optimize processes and to achieve the given results. Techniques such as financial analysis methods (e.g. performance measurement), 'unsupervised' analyses (cluster analysis, text and data mining of social media channels), 'supervised' analyses (process mining, optimization, probability calculations), regressions (linear, univariate / multivariate) and other statistical methods are applied to generate decision-relevant management information by answering the following questions (Appelbaum et al., 2017; ICV, 2016):

- What happened last season?
- Why have tour revenues been down to nearly zero and why did bike rentals increase so immensely?

- What will happen in the next season, particularly regarding Covid-19?
- What is the optimal solution to plan for the next season?

Whereas for the first two questions, deviation (descriptive) and cause and effect (diagnostic) analyses are relevant, for the last two questions forecasts (predictive analytics combined with value driver trees – i.e. prescriptive) are used to determine demand on tours and bike rentals for the next season in accordance with strategic goals. SbB will grow by establishing new markets in cooperation with partner firms.

Second, the supply chain management should be based on a digitized, largely automated and integrated information and communication technology infrastructure to plan for seasons successfully. Partners providing input (suppliers), partners participating in actual production of services (generating throughput) and finally partners receiving or consuming output (customers of tourism products) should be connected to/with a company's supply chain (Horváth et al., 2020). So-called ERP (Enterprise-Resource-Planning) systems can help large size corporates such as TUI or AIDA to connect partner platforms and to manage resources, working capital as well as customer relationships from one system. In the case of SbB, a small start-up enterprise with few employees and only a handful of tourism products, such a sophisticated system would be overstated and most functions would not be exploited in their fullest extent.

> SbB's required capacities that need to be planned continuously mostly comprise human capital in the form of guides. Bicycles represent an initial capital resource, being purchased once and thus ready for use. The utilization of both guides and bikes depends on demand which requires continuous and accurate forecasting. For guides, a planning tool would be useful; so far it is only half-digitized using Excel to ensure some means of planning and WhatsApp for communication with and coordination of staff. For bikes, booking a simple shop system (WordPress plug-in) has been applied so far. Bookings coming in via telephone or e-mail are still challenging as they induce lots of manual effort along with the risk of getting lost if standard procedures (using paper based processes) are not followed. Sales staff sensitively prompt customers to use the online booking systems to prevent overbooking and save resources, mostly in the form of manual effort (e.g. for taking down payment information and carrying out the payment process manually). The implementation of a fully digitized booking system integrated into the online shop solution is a priority for SbB within the next year.

An integrated and cloud-based platform eliminates interfaces, assures a smooth internal and external communication, generates less paper and manual work and thus reduces handling costs. However, the integration of different systems (e.g. booking, payment, customer relationship management/CRM) is linked to investment costs and/or license fees or other regular payments/disbursements depending on the contract agreements. To ensure profitability the transaction costs and commission fees for software such as planning solutions should be considered in calculations of the profit margin and prices.

> For instance, SbB pays 20% of the revenue to its distribution platforms. Paypal charged 2,49% (in 2021) plus 0.35 Euros per transaction.

For this reason, the investment in search engine optimization (SEO) over and above search engine advertising (SEA, e.g. utilising digital techniques such as banners and pop-ups) can be beneficial for advertising. With SEO, a high online search listing is guaranteed compared to competitors and thus potential customers can find the business easily. SEO is preferable to SEA as it is more sustainable in the sense that it optimizes listings in an organic manner.

> SbB realizes SEO by using WordPress plugins for the homepage to analyse its search engine performance and push its online (i.e. Google) listing. Campaigns are regularly initiated on Facebook and 'Google my business' to promote events. SEA as well as Facebook ads are rather applied for pushing demand in more temporally challenging times and in unexpected situations.

To sum up, digitization is indispensable for supporting management decision-making, for coordinating the entire supply chain and consequently steering a business successfully.

Summary

In this chapter we have examined how value chain management, in particular capacity (staff and resource) planning, supply chain contracting and building up reputational capital through social media marketing all support planning for seasonal variations in supply and demand and how crucial the role of digitization is in these processes.

Self-reflection questions

With application to any tourism related business or organisation with which you are familiar:

1. How does the value chain of a tourism business differ from that of a firm producing goods?
2. What are the key considerations for capacity planning for Stuttgart by Bike (SbB)?
3. What are the key considerations for successful supply chain management?
4. Taking SbB as an example, what measures might a seasonal tourism business take to build up its reputational capital?

References

AIDA (2020) https://www.aida.de/aida-cruises/presse/pressearchiv/newsdetails.24494/article/aida-cruises-sagt-reisen-bis-31-mai-ab-und-bietet-bonus-auf-reiseguthaben.html , 8 Apr, (Accessed 30 Nov 2020).

Appelbaum, D., Kogan A., Vasarhelyi, M. & Yan, Z. (2017) Impact of business analytics and enterprise systems on managerial accounting, *International Journal of Accounting Information Systems*, **25**, 29–44.

Bormann, P.M. (2018) *Affiliate-Marketing: Steuerung des Klickpfads im Rahmen einer Mehrkanalstrategie*, 1st edn, Wiesbaden: Springer Gabler.

Davenport, T. & Harris, J. (2007) *Competing on Analytics - the new science of winning*, 6th edn., Boston: Harvard Business Press.

Davenport, T. Harris, J. & Morison, R. (2010) *Analytics at Work: Smarter decisions, better results*, Boston: Harvard Business Press.

Franken, S. (2016) *Führen in der Arbeitswelt der Zukunft: Instrumente, Techniken und Best-Practice-Beispiele*, 1st edn, Wiesbaden: Springer.

Gartner (n.d.) Digitalization, https://www.gartner.com/en/information-technology/glossary/digitalization (n.d.), (Accessed 20 Nov 2020).

Horváth, P., Gleich, R. & Seiter, M. (2020) *Controlling*, 14th edn, München: Vahlen.

Internationaler Controller Verein e. V. (ICV) (2016) *Business Analytics Der Weg zur datengetriebenen Unternehmenssteuerung*, Wörthersee: ICV.

Kreutzer, R.T., Neugebauer, T. & Pattloch, A. (2017) *Digital Business Leadership*, Wiesbaden: Springer.

Krishnan, H. & Winter, R.A. (2011) The economic foundations of supply chain contracting, *Foundations and Trends in Technology, Information and Operations Management*, **5** (3–4), 147–309.

Lippold, D. (2017) *Marktorientierte Unternehmensführung und Digitalisierung: Management im digitalen Wandel*, 1st edn, Berlin, Boston: de Gruyter.

Maleri, R. (1997) *Grundzüge der Dienstleistungsproduktion*, 4th edn, Berlin, Heidelberg, New York: Springer.

Maleri, R. & Frietzsche, U. (2008) *Grundlagen der Dienstleistungsproduktion*, 5th edn, Berlin, Heidelberg, New York: Springer.

Meffert, H. (2000) *Marketing*, 9th edn, Münster: Springer Gabler.

Meffert, H. & Bruhn, M. (2000) *Dienstleistungsmarketing*, 3rd edn, Münster, Basel: Springer Gabler.

Pasban, M. (2016) A review of the role of human capital in the organization, *Procedia - Social and Behavioral Sciences*, **230**, 249 – 253.

Pasternack B.A. (2008) Optimal pricing and return policies for perishable commodities, *Marketing Science*, **27** (1), 131–132.

Petry, T. (2016) Digital leadership: Unternehmens- und Personalführung in der Digital Economy, in Petry, T. (ed.) *Digital Leadership. Erfolgreiches Führen in Zeiten der Digital Economy*, Freiburg: Haufe-Lexware, 21–82.

Porter, M.E. (1980) *Competitive Strategy: Techniques for Analyzing Industries and Competitors*, New York: Free Press.

Porter, M.E. (1985) *The Competitive Advantage: Creating and Sustaining Superior Performance*, New York: Free Press.

Sethi, J.A. (2017) Service marketing: An overview, in Sood, T. (ed.), *Strategic Marketing Management and Tactics in the Service Industry*, Hershey, PA, USA: IGI Global, 1-14.

Von Dörnberg, A., Freyer, W. & Sülberg W. (2018) *Reiseveranstalter- und Reisevertriebs-Management: Funktionen – Strukturen – Prozesse*, 2nd edn, Berlin, Boston: de Gruyter.

Witt, F.-J. (2003) *Dienstleistungscontrolling*, 1st edn, München: Vahlen.

Zikopoulos, P.C., deRoos, D., Parasuraman, K., Deutsch, T., Corrigan, D. & Giles, J. (2013) *Harness to Power of Big Data: The IBM Big Data Platform*, New York: McGraw-Hill.

Part 4:
Covid and Post-Covid: Temporality Futures

The final part of this book turns its attention to some key trends, both external to and enshrined within tourism, which are impacting on the nature of the relationship between temporality and tourism.

The first contribution, in Chapter 16, focuses on a number of externalities that are shaping the sectors that collectively comprise tourism. The authors consider the role of a range of mega-trends and whether they are responsible for re-shaping tourism from a temporal perspective. Such trends include structural factors (economic, social, demographic and institutional) and technological advances, which, in addition to the inherent challenges arising from climate change, are exerting influence on travel patterns, decision-making and shaping the future resource base on which destinations' appeal is based. The authors note how technological advances are re-shaping the structure of tourism service intermediation, itself so powerful on shaping consumers' demand patterns. The authors argue that the tourist value chain is effectively being redefined insofar as access to real-time data increasingly facilitates travel decisions, while enabling destinations and businesses to become better prepared to deal with temporal demand variations.

The combination of social megatrends with technological innovation has also impacted on the scale of tourism in many parts of the world. The phenomenon of overtourism was, prior to the Coronavirus pandemic, arguably one of the most challenging issues facing tourism. The resumption of travel and full-scale 'reopening' of destinations and attractions since the relaxation of Covid controls has brought the issue back to the top of the agenda in many places. This is the theme of Chapter 17, which examines and critiques the relationships between overtourism, sustainability and seasonality. The essence of the dilemma facing destinations is the balance between recovering the 'lost ground' for businesses and tax revenues from the two years of minimal tourism activity (2020-2021) and achieving meaningful sustainability of the natural, built, cultural and community resources of the place going forwards. The author provides examples of destination responses to peak season overcrowding, with a particular focus on the challenges for Venice.

Finally, a 'post-script' chapter provides an assessment of the impacts and implications of the Coronavirus pandemic on temporality in tourism, followed by a review of some of the key issues emerging from various chapters. It poses the question 'is there an end to temporality?' in light of the post-Covid recovery and potential inexorable increase in tourism going forward.

16 The Growth of De-Temporalisation in Tourism

Alisha Ali and Philip Murray

Learning outcomes

The purposes of this chapter are:

1. To examine if there is a place for the de-temporalisation of tourism
2. To discuss the mega trends de-temporalising tourism
3. To examine the role of technology in this de-temporalisation and the future implications
4. To consider future developments which will continue to drive de-temporalisation of tourism

Introduction

Tourism is more than physical travel to another place. For many, tourism offers educational, cultural, spiritual and life affirming benefits going beyond the generic (business or leisure) motivations for travel. Individual values also influence and motivate the type of tourism experience desired. Over time tourists' motivations, knowledge, desire and sophistication will change but the notion of seasonality will retain a strong influence on when, how and where people travel. Many tourist economies use this seasonal influence to efficiently manage capacity and maximise revenue, however this is only one aspect of tourism's temporal relationship. Dynamic megatrends and tourism's digital revolution are challenging this traditional notion of seasonality by creating temporal shifts.

Seasonality is often considered as a temporal imbalance problem (Cannas, 2012) with destinations seeking ways to balance the peaks and troughs of tourist demand to improve efficiency and profitability. This chapter focuses on de-temporalisation as it applies to seasonality. De-temporalisation of seasonality is, in effect, the resolution of the temporal imbalance problem by the smoothing of tourism demand across an entire period and / or the creation of multiple seasons in a given period. This chapter examines these key megatrends including the role of technology in de-temporalising the tourism industry.

De-temporalisation megatrends

Megatrends are large, slow forming global shifts in our political, economic, social, technological, environmental and legal settings which re-shape the way we live and work. These megatrends, also known as macroeconomic and geostrategic forces (PWC, 2016) can irreversibly shape and reshape our tourism operating environment with a direct influence on tourism demand. Structural, fruition and climate and unforeseen factors have been identified as important megatrends to investigate for the tourism industry (Senbeto & Hon, 2019). These three trends are discussed below, which encompasses demographic change, an emerging middle class, resource depletion and climate change, all of which have been identified as having a significant impact on the tourism industry (Horwath HTL, 2015; OECD, 2018; United Nations, 2020).

Structural factors

Structural influences on tourism movement relate to economic, social and policy characteristics which affect an individual's ability and propensity to engage with tourism and travel. The first influential megatrend is demographic, signifying social changes which are forecast to take place over the next 30 years and are likely to have a significant influence on when, how often and where people holiday. In effect, this will promote de-temporalisation by changing the meaning of seasonality for destinations. The world's population is expected to grow to almost 10 billion people by 2050 with the largest growing category of the population being the over 60s, who will account for over 20% of the world's population by the middle of the century (United Nations, 2019). Older people travel in different patterns to their younger counterparts, with data for the EU 27 (Eurostat, 2018) highlighting that the 65+ age group travelled more frequently, for longer periods and were more likely to travel in traditionally 'shoulder' periods, effectively extending the traditional summer and winter seasons preferred by younger tourists. An aging population offers tourism business opportunities to extend their busy seasons by catering to older travellers. It might also mean that selective targeting of older travellers might create new markets outside of established travel periods.

The world is also getting wealthier. It is predicted that the burgeoning middle classes of India, China and Brazil will account for over 50% of the world's economic output by 2050, with North America and Europe contributing slightly less than one third in total (United Nations, 2019). Traditionally, an increasing middle class has more disposable income and is more inclined to travel as part of their leisure consumption, initially in their home country but subsequently internationally (Zeng & Go, 2013), This opens a wealth of new opportunities to cater to different cultures and peoples who are likely willing and able to travel at different times to traditional 'western' seasons due to different religious or cultural observances or school holidays: for example, the emergence of halal tourism, a segment whose travel expenses are expected to exceed US$220bn by 2030 (Vargas-Sánchez & Moral-Moral, 2019). Halal tourism products and services are adapted to conform with Sharia law, illustrated by destinations which will offer separate male/female facilities, a lack of alcohol or pork products and the announcements of prayer times, so that tourists may continue to practice their religious observances while away from home.

Another prominent group is the millennial travellers who make up the largest segment of travel clients and tend to have different preferences than previous generations. For example, research on European millennial travellers (Ketter, 2020) indicated that they value technology-enabled, personalised, 'off the beaten track' tourism, with a focus on alternative accommodation. Responding to these preferences and designing tourism products to best satisfy these needs on a year-round basis will become a key factor in the de-temporalisation of the tourism product.

Fruition and climate factors

Both fruition factors (the realisation of travelling plans) and climate (weather) are likely to be strongly influenced by climate change and resource depletion. Fruition factors influence travel that is goal-orientated, like visiting friends and family, business or leisure; while the climate of a destination has long been a significant influence in tourism motivation. Increased awareness of the impacts of tourism on sustainability will likely influence destination choice for travellers in the future. This provides both challenges and opportunities for destinations as they can seek to balance sustainability considerations with the need to generate revenue. This was seen in the recently overturned plans to close Komodo Island (the only place to see the Komodo dragon in the wild) to tourists due to economic concerns from the heavily tourism dependant local economy. Tourists of the future are much more likely to consider the lasting impact of their travel choices and the carbon footprint involved. This may fundamentally change where and when people travel as they seek a more benign tourism experience. Destinations will also have to respond by limiting capacity, changing the tourism offer or closing entirely, which will impact on the destination's temporality.

Climate change will bring increased temperatures, sea levels and instances of extreme weather which will impact on the viability of destinations and their productive seasonal capacity. Maya Bay in Thailand represents an example of both the impacts of climate change and negative impacts of tourism. Previously receiving up to 5,000 visitors a day before closing to the public in 2018 in an effort to rejuvenate its decimated coral reefs, the re-opening of the destination was delayed due to a heatwave in 2019 which damaged some of the newly planted coral. The destination was expected to re-open in 2021 and limit its visitors to approximately 1,200 per day (BBC News, 2019).

Unforeseen factors

Unforeseen occurrences can dramatically impact on the seasonal demand of a destination and are by their nature almost impossible to plan for. Extreme weather events, natural disasters and terrorist attacks can have both an immediate impact on tourism in terms of an instant decrease in demand but may also further erode confidence in a destination's image having a more long-lasting impact. The close relationships between tourism demand

and economic cycles and the susceptibility of tourism to global shocks are well established. A robust crisis management and marketing response is required to rehabilitate destinations after an external shock. Often time a destination's response should be supported by policy intervention on a national or transnational level. Many of the megatrends may contribute to unforeseen occurrences which influence tourism demand and seasonality. Most notable in its potential to influence is climate change, but as the recent Covid-19 pandemic has demonstrated we may not always be able to predict sources or impacts of these influences. IATA (2020) predict an $118.5 billion net loss for the airline sector in 2020 as aviation demand has been obliterated by global lockdowns. The impact of the crisis is set to be three-times more damaging to the sector than the 2009 global financial crisis.

Understanding the influence of megatrends and developing contingency responses to their consequences are, as suggested by Fayos-Solà and Cooper, (2018) part of an evidence and knowledge-based approach to tourism which *"helps to predict and so strategically manage the future, rather than allowing tourism to become a straw in the wind of technological, and institutional change"*. Change is inevitable and can have a significant influence on how and when tourists choose to travel. Destinations too must change to best serve new markets, new customer preferences and to provide new experiences for customers. De-temporalisation of tourism will be a direct result of the changes in tourist preference and travel motivations and will allow the sector to maximise capacity and allow destinations to benefit from stable demand and consistent revenues. An additional significant factor in the growth of de-temporalisation of tourism is the role and influence of technology. This will now be examined in detail in the following section.

Technology and de-temporalisation

Tourism's digital revolution is influencing temporality because it has transformed the structure of the industry, creating a paradigm shift in marketing of destinations. The fragmented and information-intensive nature of the industry makes it more responsive to the benefits of technology (Buhalis & Deimezi, 2004) and in essence temporality. The ubiquity of the Internet has transformed technology from a support tool in the form of proprietary global distribution systems, reservations and booking systems in the latter half of the 20th century to now being embedded in the entire tourism ecosystem. This married with big data and the Internet of Things is altering the tourism system and contributing to de-temporalisation. Figure 16.1 provides a timeline of the development and use of technology in tourism.

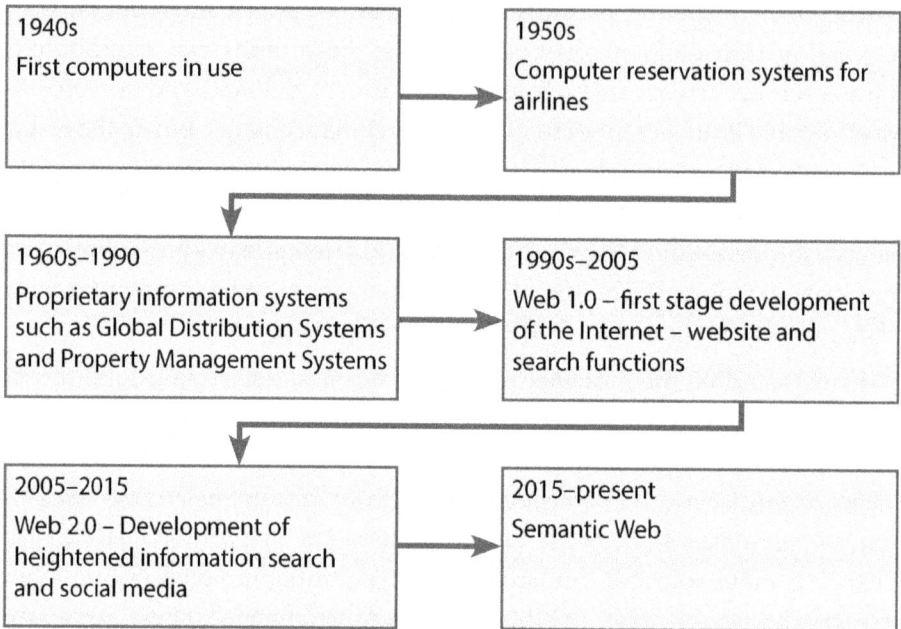

Figure 16.1: Development of technology use in tourism

Disintermediation and re-intermediation

Notions of de-temporalisation started with Web 1.0 and the disintermediation of the tourism industry. Distribution has been one of the most impacted areas in tourism due to technology (Cetin et al., 2016). The tourism distribution channel refers to the suppliers and intermediaries that enable the sale and delivery of travel services to tourists. Via the Internet, tourism businesses were able to have websites and communicate directly with the tourists without the reliance on and restrictions imposed by the middle person such as travel agents and tour operators. Disintermediation led to a period where customers booked directly through the supplier and searched for information online anytime, and suppliers such as hotels, rental car agencies and tour companies invested heavily in their websites and search engine optimisation.

As the Internet developed from Web 1.0 to Web 2.0, it became a marketing and sales medium which fostered deeper customer interactions (Cetin et al., 2016). This change in the internet architecture fostered reintermediation of the industry with the growth of online travel agencies (OTAs), online third-party intermediaries which virtually connect tourists and suppliers. OTAs became dominant because they shifted the power from suppliers to themselves.

The Internet was now the customer's first choice in tourism information search (Law et al., 2014) as the effort to search for information shifted from the business and intermediaries to the consumer. Tourists compared information from different sites before making a final booking (Law & Huang, 2006) and it was estimated that they would visit more than 20 websites before making a final purchase (Thakran & Verma, 2013). However, this comparison became time consuming and led to information overload (Law et al., 2007) which in turn created the opportunity for the development of meta search engines such as Skyscanner and Kayak, which collect information from different sites into one display. Alongside this development was the sophistication of search engines. Internet traffic is now being directed via these search engines as opposed to a direct website, adding to the re-intermediation of the industry (tom Dieck et al., 2018). OTAs and metasearch sites allowed a business to be presented over more than one distribution channel. They aggregated holiday packages and allowed tourists to dynamically package their own holidays.

OTAs have contributed to de-temporality because they served the dual purpose of being business-to-business and business-to-customer sites, thereby extending the reach and market exposure of suppliers. Here the real potential of the internet was seen for tourism, and in essence the rise of 'temporal demystifiers'. In other words, whilst seasonality still played a part in the tourist offer, opportunities were emerging to create off-season offers to the tourists, channelled through the change in online distribution of the tourism product and the growth of direct connections through social media. Price is the key stimulant for travel online purchase (O'Connor & Murphy, 2008). OTAs and meta search sites allowed tourists to search and gather travel intelligence to find the best travel products suited to their needs. It allowed suppliers to showcase different types of events and activities, enticing tourists to book out of season activities, hence shifting the notion of time for tourists. These OTAs have made trips more accessible to the customers based on their travel preferences and requirements. Additionally, trip planning evolves on having the availability of attractive travel choices based on what customers would like to see and do in the region.

Social media and the handset culture

Social media (blogs, social networking sites, content communities etc.) are now an important part of the tourism landscape as they play a critical role in tourist information search, decision making, marketing and customer interaction (Law et al., 2014). They have opened pathways of information exchange for the tourist, enabling them to reach the right communicator. Such co-creation is enabling tourists to create their own value and experiences and change the service management process.

The growth of mobile technologies is fuelling these social media conversations. Tourists are now the connected and conversation customer due to the ubiquity of mobile phones. They are also content creators, because they are engaged in co-creating their experiences and assume various roles as part of this such as *"booking, (self)guiding, reviewing, sharing and marketing the destination"* (Dredge et al., 2018, p. 10). Social media has helped to reduce the intangibility of the tourism product offer by providing information to ensure they are making the right choices, maximising their value and how they spend their time. This reduction in intangibility provides a sense of security for tourists to travel 'out of season' and consume new experiences. Tourists now have choice, flexibility and connectedness to engage in service transactions, challenging the traditional notion of when and how they should travel. From the supplier perspective, social medial has allowed them to be marketed on a global stage instantaneously sharing experiences which the tourist can participate in, exploiting new markets and changing the tourism demand patterns.

Disruptive innovations

Technology has permanently altered the structure of the industry and created new business models which has impacted on the temporality of the industry. One such example is the sharing economy, which developed because technology enabled the creation of digital platforms to temporarily share resources for an economic return, which could be monetary or non-monetary. Well-known examples are AirBnB and EatWith. These businesses are connecting customers directly with hosts as tourists were searching for cheaper and more authentic local experiences and offering alternatives to the traditional hotel and food and beverage and tours offerings.

The Internet has also enabled the growth of low-cost airlines which is opening access to new markets both domestically and internationally. Markets which were once closed have now become open and easier to access. For example, prior to the Covid-19 pandemic, RyanAir launched direct flights from Jordan to Europe. Such disruptions are temporal de-mystifiers as they are enabling travellers to have more choice and flexibility.

Improving accessibility

Technology has also enabled accessible and inclusive travel allowing those with special requirements to still experience the tourism offer. Tourists with disability requirements need to have information specific to their accessibility needs before any trip is booked (Michopoulou & Buhalis, 2013). Using websites, social media and engagement with virtual reality, tourists can be provided with this information pre-trip to allow them to make informed decisions. Technology has created opportunities for travel for populations which may have been previously excluded. This supports the de-temporalisation of the industry as these tourists may travel during off-peak seasons based on their needs. At the destination, technology can support visitors who have visual, mobility, auditory and cognitive impairments by helping them to navigate physical and service barriers (Michopoulou & Buhalis, 2013).

Future developments

It is difficult to predict the future megatrends which will challenge the tourism industry, however, these will be shaped by our social, political, economic and technological environments. New technological advancements in artificial intelligence (AI), augmented reality, virtual and mixed reality, Internet of Things, haptics, blockchains, wearable devices, smartphones, Web 4.0, 5G mobile networks, gamification (Buhalis et al., 2019; Tussyadiah et al., 2018) will shape the future of temporality. These technologies are changing the tourist behaviour and redefining the tourism value chain. This will enable smart tourist destinations to arise out of need for the *"infostructure of co-creating value"* (Gretzel et al., 2015). The development of smart tourism destinations fuelled by the interconnectivity and interoperability of technology will transform the tourism system to provide innovative services through co-creation, personalisation and context awareness for stakeholder value

(Buhalis, 2020). In essence, via technology, all the players in the tourism ecosystem are connected, which allow them to engage with real-time data.

AI is already advancing at a rapid pace with it being adopted by OTAs, airlines and tour operators. It has been estimated that up to 70% of future online tourist transactions will be processed by AIs (Kazandzhieva & Santana, 2019). AI has the capability to learn new information quickly to solve problems, resulting in faster decisions over a wider reach, enabling more optimised and personalised services to tourists.

In real time, virtual reality (VR) simulates the user senses by enabling visualisation, navigation and interaction in a computer-generated 3D environment (Beck et al., 2019). Augmented Reality (AR) is different because the real-world environment is overlaid with computer-generated information through a device (Jung et al., 2016). AR/VR benefits the tourism industry through educating and entertaining the tourists and can be used creatively to market destinations and deliver innovative experiences. Virtual reality has led to tourists having a sense of being (Tussyadiah et al., 2018) whilst AR can increase place attachment (Oleksy & Wnuk 2017). This is important for temporality because it offers the tourist different ways to experience the destination. It provides opportunities for tourism marketers to communicate their offer to different target markets based on their preferences and this can be located within or outside of defined 'seasons'. It can also educate them about the right time for them to visit a destination.

Finally, blockchain consist of records of transactions known as blocks which are attached to each other. This enables large databases to be created. Blockchain holds promise of revolutionising tourism because the security features create trust, which is important in online booking. This enables more secure digital identification leading to quicker service delivery, more transparent transactions and more secure online data storage. Blockchain may lead to another wave of disintermediation because its security features will enable more trustworthy peer-to-peer transactions eliminating the need for intermediaries such as OTAs (Kazandzhieva & Santana, 2019). Currently, intermediaries provide this trust to the consumers by checking and verifying the suppliers. With blockchain, this will no longer be required. The TUI Group, CoolCousin, UK, Travelchain, Russia and Webjet, Australia have already started to engage with blockchain and new distribution platforms such as Winding Tree are emerging. Blockchain can support the temporal nature of the industry due to the trust created for tourists. This breaks the reliance on third parties such as OTAs, allowing tourism to dynamically package and book holidays suited to their convenience. These disruptive technologies create agility in the tourism system *"empowering the co-creation of value for all stakeholders"* (Buhalis, 2020, 269). This real time interaction reinforces the de-temporality of the tourism system.

Summary

The mega trends and the continuous innovation in technology will undeniably create disruptions in the tourism industry and will facilitate new tourism experiences. To embrace these opportunities, tourism destinations and business need capable leadership to drive these changes forward and understand how such changes can lead to de-temporalising the tourism industry. The fourth industrial revolution enabling the development of smart tourism provides destinations with the tools to understand, share, govern and innovate. This creates a more connected and integrated destination facilitating sustainability and destination resilience.

In understanding the megatrends and the influence of technology on tourism temporality, it is important to consider the challenges. Historically, destination management organisations have been slow to recognise the changing marketplace and implement technology due to finances, knowledge and their operational structure. With technology, one must consider privacy and security issues, digital inclusivity, system failure, change in the service experience and the human experience, cost and fitness for purpose. The dynamics of these developments requires tourism stakeholders to respond smartly to these changes. They need to understand tourist behaviour, specify their competitive advantage, be flexible and adaptable to the changes in the environment. Such knowledge will allow them to use advances in technology as the tools of detemporalisation to help them achieve differentiation and a competitive advantage.

Vignette: Destination Wow

It is autumn 2071 in Destination Wow. It is hotter than usual due to worldwide changes in weather patterns, lengthening their traditional summer season. Wow has invested in a destination dashboard where it uses technology to monitor and plan for its longevity. With the extended season, Wow uses this system to predict future room bookings by interrogating data from previous bookings on search engines, meta search sites, OTAs and other booking platforms. Wow can now manage the existing demand and determine if there is a need to shift if capacity is reached. This prompts the destination to consider the experience-based activities to keep customers motivated to travel to the destination. Using artificial intelligence (AI), the destination dashboard creates customised offers based on tourist behaviours utilising previous travel history or known preferences from data mining. This is promoted through hyper-personalised communication to the connected tourists whom Destination Wow is seeking to attract.

Destination Wow is also benefitting from new markets with the growing affluence of the middle class in the Pacific Rim. In the past 50 years, the destination has seen more diverse travellers and a growing market of younger and active older tourists. This resulted from faster air travel with hypersonic airplanes, more direct access due to budget airlines and hassle-free travel due to digital visas and border checks. Technology has been used to upgrade its accommodation stock with AIs, robots, high-speed internet access and environmental management information systems. Ground transport systems are now intelligent, providing real-time destination travel information and flying taxis. Sites and attractions have created meaningful interpretation experiences through AR, VR and holographics. Wow wants to be the destination of the future and has invested in technology to develop a resilient, year-round tourism offer, ensuring it is responding to customer demands and managing shocks in the tourism system.

Self-reflection questions

1. What are megatrends and how will they impact the temporality of the tourism industry?
2. How has technology shaped the disintermediation and re-intermediation of the tourism industry?
3. In what ways has technology supported the de-temporalisation of the tourism industry?
4. What technologies can be used by destination planners to create more even demand and supply in the tourism system?
5. Future megatrends are indicating that tourists will continue to seek more personalised, tailor-made experiences that buck established temporal patterns of visitation. How can the digital revolution in tourism support this?

References

BBC News (2019) Thailand: Tropical bay from 'The Beach' to close until 2021. https://www.bbc.co.uk/news/world-asia-48222627.

Buhalis, D. (2020). Technology in tourism - from information communication technologies to eTourism and smart tourism towards ambient intelligence tourism: a perspective article. *Tourism Review*, **75**(1), 267-272.

Buhalis, D. and Deimezi, O. (2004), E-tourism developments in Greece: Information communication technologies adoption for the strategic management of the Greek tourism industry, *Tourism and Hospitality Research*, 5 (2), 103-130

Buhalis, D., Harwood, T., Bogicevic, V., Viglia, G., Beldona, S. & Hofacker, C. (2019). Technological disruptions in services: Lessons from tourism and hospitality. *Journal of Service Management*, **30**(4), 484-506.

Beck. J., Rainoldi, M., & Egger, R. (2019). Virtual reality in tourism: A state-of-the-art review. *Tourism Review*, **74**, 586-612.

Cannas, R. (2012) An overview of tourism seasonality: key concepts and policies, *Journal of Tourism, Culture and Territorial Development*, 3(5), 40–58.

Cetin, G., Cifci, A.M., Dincer, F.I. & Fuch, M. (2016). Coping with reintermediation: The case of SMHEs. *Information Technology and Tourism*, **16**, 375–392.

Dredge, D., Phi, G., Mahadevan, R., Meehan, E. & Popescu, E. S. (2018).

Digitalisation in Tourism: In-depth analysis of challenges and opportunities. Executive Agency for Small and Medium-sized Enterprises (EASME), European Commission. https://ec.europa.eu/docsroom/documents/33163/attachments/1/translations/en/renditions/native

Eurostat (2018) *Tourism Trends and Ageing*: https://ec.europa.eu/eurostat/statistics-explained/index.php?title=Talk:Tourism_trends_and_ageing

Fayos-Solà, E. & Cooper, C. (2018) Conclusion: The future of tourism-innovation for inclusive sustainable development, in *The Future of Tourism: Innovation and Sustainability*. Springer International, pp. 325–337. doi: 10.1007/978-3-319-89941-1_18.

Gretzel, U., Sigala, M., Xiang, Z. & Koo, C. (2015). Smart tourism: foundations and developments. *Electronic Markets,* **25**(3), 179-188.

IATA, 2020. Deep losses continue into 2021. https://www.iata.org/en/pressroom/pr/2020-11-24-01/ (Accessed 31 March 2021)

Horwath HTL (2015) Tourism megatrends 10 things you need to know about the future of tourism. http://corporate.cms-horwathhtl.com/wp-content/uploads/sites/2/2015/12/Tourism-Mega-Trends4.pdf

Jung, T., tom Dieck, M.C., Lee, H. & Chung, N. (2016). Effects of virtual reality and augmented reality on visitor experiences in museum. In: Inversini, A. and Schegg, R. (Eds), *Information and Communication Technologies in Tourism 2016,* Proceedings of the International Conference Bilbao, Spain, 2-5 February, New York: Springer, pp 621-635.

Kazandzhieva, V. & Santana, H. (2019). E-tourism: Definition, development and conceptual framework. *Tourism,* **67**(4), 332-350.

Ketter, E. (2020) Millennial travel: tourism micro-trends of European Generation Y, *Journal of Tourism Futures* 7(2). doi: 10.1108/JTF-10-2019-0106.

Law, R., Buhalis, D. & Cobanoglu, C. (2014). Progress on information and communication technologies in hospitality and tourism. *International Journal of Contemporary Hospitality Management,* **26**(5), 727-750.

Law, R., Chan, I. & Goh, C. (2007). Where to find the lowest hotel room rates on the Internet? The case of Hong Kong. *International Journal of Contemporary Hospitality Management,* **19**(6), 495-506.

Law, R. & Huang, T. (2006). How do travellers find their travel and hotel websites? *Asia Pacific Journal of Tourism Research,* **11**(3), 239-246.

Li, X.R., Lai, C., Harrill, R., Kline, S. and Wang, L. (2011). When east meets west: An exploratory study on Chinese outbound tourists' travel expectations, *Tourism Management,* 32(4), 741-749.

Michopoulou, E. & Buhalis, D. (2013). Information provision for challenging markets: The case of the accessibility requiring market in the context of tourism. *Information & Management*, **50**(5), 229-239.

O'Connor, P. & Murphy, J. (2008). Hotel yield management practices across multiple electronic distribution channels. *Information Technology & Tourism*, **10**(2), 161-172.

OECD (2018) *OECD Tourism Trends and Policies 2018*. doi: 10.1787/tour-2018-en.

Oleksy, T. & Wnuk, A. (2017). Catch them all and increase your place attachment! the role of location-based augmented reality games in changing people–place relations, *Computers in Human Behaviour*, **76**, 3–8.

PWC (2016) Five megatrends and their implications for global defense & security. https://www.pwc.com/gx/en/government-public-services/assets/five-megatrends-implications.pdf .

Senbeto, D. L. & Hon, A. H. Y. (2019) A dualistic model of tourism seasonality: approach–avoidance and regulatory focus theories, *Journal of Hospitality and Tourism Research*, **43**(5), 734–753. doi: 10.1177/1096348019828446.

Thakran, K. & Verma, R. (2013). The emergence of hybrid online distribution channels in travel, tourism and hospitality. *Cornell Hospitality Quarterly*, **54**(3), 240-247.

tom Dieck, C.M., Fountoulaki, P. & Jung, T. (2018). Tourism distribution channels in European island destination. *International Journal of Contemporary Hospitality Management*, **30**(1), 326-342.

Tussyadiah, I.P., Jung, T.H. & tom Dieck, M.C. (2018). Embodiment of wearable augmented reality technology in tourism experiences. *Journal of Travel Research*, **57**(5), 597-611.

United Nations (2019) *World Population Prospects 2019 Highlights*. https://population.un.org/wpp/Publications/Files/WPP2019_Highlights.pdf

United Nations (2020) *Report of the UN Economist Network for the UN 75th Anniversary: Shaping the Trends of Our Time*. https://www.un.org/development/desa/publications/wp-content/uploads/sites/10/2020/09/20-124-UNEN-75Report-2-1.pdf (Accessed: 11 January 2021).

Vargas-Sánchez, A. & Moral-Moral, M. (2019) Halal tourism: state of the art, *Tourism Review* **74**(3), 385–399. doi: 10.1108/TR-01-2018-0015.

Zeng, G. & Go, F. (2013) Evolution of middle-class Chinese outbound travel preferences: an international perspective, *Tourism Economics*, **19**(2), 231–243. doi: 10.5367/te.2013.0202.

17 Seasonality and Overtourism

Richard Butler

Learning outcomes

This chapter will provide you with:

1. An understanding of the concept and definition of 'overtourism'.
2. The inter-relationships between seasonality and overtourism and the factors inherent to that inter-relationship.
3. An appreciation of how overtourism and seasonality in tourism can be open to mitigation or prevention.

Introduction and context

Writing at a time when Covid-19 has been prevalent throughout the world, to discuss issues such as seasonality and overtourism may seem both inappropriate and insensitive. Many tourist destinations and most elements of the tourism industry have, during 2020 and 2021, severely suffered economically from an absence of tourists rather than an imbalance or surfeit of visitors. The concerns about overtourism in particular, expressed so strongly in the last years of the second decade of the 21st century (see for example, Dodds & Butler, 2019; Milano et al., 2019), now seem strangely irrelevant and unmemorable. However, it is likely that what are viewed as major difficulties in tourism, both seasonality and overtourism, are almost certain to return to many destinations in the not too distant future. The fact that so many potential tourists were denied the opportunity to travel, particularly to foreign destinations, for almost the whole of 2020 and much of 2021, means that there is a great deal of latent demand forcibly pent up in most, if not all, of the traditional major origin countries.

At the time of writing the Covid pandemic has made overtourism and seasonality almost irrelevant, but when Covid-related restrictions are reduced or removed, it is hard to imagine that tourist numbers will not rise rapidly almost immediately, perhaps not to the previous record levels of 2019 but certainly to levels causing crowding and negative responses in more popular destinations. How quickly tourist numbers will rise and in which destinations will depend on a number of factors: the perceptions of tourists about health and safety risk and the availability of medical care; the removal of enforced quarantine at both destinations on arrival and at home on return; the removal of other government restrictions on travel; the availability of transportation services (aircraft being in the appropriate place and in service for example), of accommodation and related services, and of appropriate intermediary services to make travel possible. If constraints are removed or at least significantly reduced, crowds can be expected to return to popular sites and sights and the spectre of overtourism to reappear, despite the possibility of forces opposed to tourism in general attempting to seize what they may see as an opportunity to significantly and permanently reduce the scale of the activity (Butcher, 2020; Tourism Geographies, 2020).

In light of the above, this chapter explores the inter-relationships and debates around overtourism and seasonality, the difficulties caused by inconsistency in terminology, and the reluctance of authorities to tackle the 'wicked problem' of excess visitation. The example of Venice is used to illustrate the long-lasting nature of the problems faced by some destinations.

Overtourism

The phenomenon of overtourism, an 'excessive number of visitors', has become highly visible in tourism, both academically and more particularly in reality, during the last half decade, finding its strongest expression in a number of locations, mostly urban centres, especially in western Europe (Dodds & Butler, 2019). Places such as Venice, Barcelona, Dubrovnik, Prague, Edinburgh and Paris all have featured in both the popular press and news media sites and in academic sources (Dodds & Butler, 2019; Milano et al.,

2019). The phenomenon even gained acknowledgement from bodies such as UNWTO (2018) and Peeters et al. (2018).

In reality, an excessive level of crowding and visitation by tourists in a variety of locations across the globe was neither new nor confined to the urban centres receiving most of the media attention, a view argued by Buhalis (2020) and Dredge (2017) amongst others. Overcrowding, which is really what overtourism represents, was recognised as a potential, if not an existing problem, in recreation and tourism in the 1960s (Clawson, 1959; Darling & Eichorn 1967). But, as has become obvious over the intervening period, successful destinations have rarely, if ever, been concerned with too many visitors. Most were focused on receiving even more tourists than in previous years. Success in the context of tourism and recreation destinations has traditionally been measured in numbers of visitors, despite the reality that increasing numbers are not an automatic sign of greater economic benefit (Wilkinson, 1996), which depends on additional factors such as length of stay, expenditure per head, where such expenditure is made, and level of leakage at the destination.

The tourism industry and its formal mouthpieces such as UNWTO and WTTC have in the past denied the idea of 'excessive numbers', arguing that any problems that did exist were problems of management and lack of dispersal of visitors, rather than of numbers per se (WTTC, 2018; UNWTO, 2018). To take steps to reduce visitor numbers is not something which any Destination Management Organisation has been willing to propose, and such actions have not been widely supported by major tourism bodies to date. However, a few specific locations such as Amsterdam (Gerritsma, 2019), have announced an intention to reduce the promotion of additional tourism.

A reluctance to reduce or limit visitor numbers to parks and other natural areas can be found among some natural area management agencies who fear budget cuts because of political decisions should numbers decline (personal communication, Ontario Parks). Even highly popular protected areas such as the Galapagos Islands have experienced constantly increased tourist arrivals, in that case from 11,765 tourists in 1979 to 241,800 in 2017 (Pecot & Ricaurte-Quijano, 2019).

In rural areas, poorly conceived developments designed to attract tourists have increasingly met opposition from negatively impacted local residents (Butler, 2019) and popular locations with tourists in several countries have been slow to impose and/or implement appropriate actions to counter overuse, e.g. Boracay in Philippines (Crux & Legaspi, 2019). That overtourism has materialised should not, therefore, have come as any surprise to destination managers, tourist authorities at all levels or local residents. As well as being almost inevitable, given attitudes towards tourism development in most quarters, overtourism has been increased in its occurrence and effects by temporal factors, in particular seasonality.

Relationship between overtourism, sustainability and seasonality

That there are links between overtourism and seasonality seems indisputable. Virtually all media articles and much of the academic literature discusses overtourism in the context of it occurring during the peak tourist 'season' (whenever that may be) at specific locations, implying, if not actually stating, that at non-peak times most of these specific destinations are experiencing acceptable levels of visitation. While a few opponents of both overtourism, and tourism in general, express a desire for a reduction in all forms of tourism (Tourism Geographies, 2020), most writers on the subject are opposed specifically to what is portrayed as excessive numbers of visitors in specific locations. These tourists, by their sheer weight of numbers, have resulted in negative effects on the quality of life of both permanent residents and also visitors to those affected destinations, resulting in calls for a reduction in numbers of visitors at specific times.

The desire is thus to reduce tourist numbers to levels experienced in earlier years and/or at quieter times of the year. The significance of seasonality, or temporal fluctuations in numbers of visitors in this context, is clear. Allied with this viewpoint, emphasising the temporal dimension of seasonality, is the frequently proposed measure of dispersing tourist numbers from peak times to other times of the year to avoid excessive numbers of visitors (UNWTO, 2018). Such attitudes have also been expressed in the context of sustainable tourism, namely that reducing the seasonal nature of tourism would increase the level of sustainability of tourism. The link between the two concepts, sustainability and seasonality, is not explicit and has never been researched to a level which might justify such a viewpoint. Sustainability clearly is seen as a contrasting situation to overtourism, but exactly how and why this should be, has neither been fully explained nor properly researched.

Nevertheless, it can be argued with some justification that overtourism as a concept is not compatible with the concept of sustainable development (Mihalic, 2020). Overtourism clearly exceeds what may be regarded as the capacity of the social component at least of the triple bottom line (environmental, social and economic elements) embedded in the general concept of sustainable development, whether that be in the context of tourism or some other activity. Reducing overtourism could, therefore, be argued to represent moving towards a state of sustainability in any specific affected destination. Given that overtourism has, as noted above, been related to times of peak visitation of tourists, then reducing numbers at specific times should result in a greater level of sustainability in those destinations.

One problem with this argument is that while such a development may result in a lessening of the negative impacts of tourism in terms of the social carrying capacity of destinations, it is not necessarily compatible with improving sustainability with respect to the environmental and economic conditions. Re-allocating visitors temporally to gain a more even spread throughout the year, in effect lengthening the season, may result in increasing the environmental impacts of visitors at other, even more sensitive times of the year, for example, during the germination of plants or reproduction of wildlife. Encouraging or implementing visitors to travel to destinations at otherwise quiet times of the year, even if it were possible to ensure such visitors were those who normally came at peak times (thus reducing overtourism), might result in disturbance of residents at times when they were enjoying a reduced level or near total absence of visitors. The attitude of residents being glad to see visitors arrive at the start of the season but being equally glad to see them depart at the end of the season (as noted decades ago by Brougham and Butler, 1981) is not unique to rural areas. The 'off-season' can be both a time of relaxation from the presence of visitors as well as an opportunity to engage in a number of other activities, ranging from the maintenance of facilities to the taking of residents' own vacations (Lee et al., 2008).

Thus describing overtourism as a seasonal problem would be somewhat simplistic and in some cases barely relevant (see the case study on Venice). For example, Milano et al., (2019) makes only one reference to seasonality in their volume on overtourism. In the context of Amsterdam, Gerritsma notes *"Overcrowding is experienced chiefly in specific circumstances"* (2019: 139), a view somewhat supported by Visenti and Bertocchi who define overtourism as *"an occurrence of far too many visitors for a particular destination to absorb over a given period"* (2019:20). This also implies a temporal element in the occurrence of the phenomenon. However, in the discussion of overtourism/overcrowding in Amsterdam (Gerritsma, 2019) there is no mention of seasons or changing times of visitors, only of reallocation of visitors spatially. Specific discussion of overtourism in the context of seasonality or temporal variations is also lacking in Dodds and Butler (2019).

While other tourist destinations are not so overwhelmed as Venice for such a large part of the year, it is clear that overtourism can exist beyond what may be the peak season. This may be due more to the fact that overtourism is essentially a personal response by residents of affected destinations to the effects which they deem to be caused by excessive numbers of tourists. Such a feeling can manifest itself at any point in time depending on the circumstances and situation of the residents and their interaction with tourists. In some situations, the presence of tourists, even in small numbers that are well below the numbers experienced in peak season, can be disruptive and unwelcome if they interfere with actions of residents, whether those be related to economic activity or pleasure. Thus interference by tourists with funerals in Orkney has been related both to criticisms of overtourism and also cruise visitors in those islands (Horne, 2016) although it is not clear if numbers of visitors there could really be described as overtourism, nor if the type of problem noted is confined to one or two specific locations in the islands or widespread throughout.

Many problems relate more to the varying and imprecise definitions of overtourism than to the time of the year in which numbers of visitors are excessive. For example, there is no agreement over the appropriate ratio of visitors to residents in a destination that characterises overtourism, nor over whether all visitors or only some, for example cruise passengers, constitute 'overtourists'. Nor is it clear whether overtourism has to occur throughout a destination's 'season' or can appear at what might simply be regular peak visitor periods, namely weekends, holidays and special occasions such as festivals, carnivals, ceremonies or events such as 'Golden Week' in China. Such occasions are always anticipated to be busier than normal and resident attitudes and tolerance may well have become adjusted to take these occurrences into account.

One might anticipate that destinations, particularly those close to large population centres, may well accept high levels of crowding during an Easter weekend or other national and local holiday periods, reasonably secure in their belief that at the end of that period, visitor numbers will return to a more acceptable level. The concerns expressed about overtourism are generally couched in terms of excessive numbers over an extended period of time, 'the summer' for example, or 'the season', implying that resident complaints reflect as much the duration of high numbers, as they do the levels themselves. In other words, it is the cumulative effect of large numbers of visitors over a long period that is at the root of the dissatisfaction, while extremely busy periods of short duration are regarded as a normal annoying element of tourism in that community but do not warrant being described as overtourism.

Other issues

This absence of attention to seasonal variations in the presence or absence of overtourism in busy tourist destinations suggests a number of other factors are involved. One may be that the feeling of overtourism may reflect the level, type, and frequency of contact between residents and visitors. Brougham and Butler (1981) found contact to be the major single factor in explaining attitudes of residents towards tourists in their study. McFarland noted that residents of Niagara Falls altered their patterns of visitation to the town centre to avoid weekends, particularly, but not exclusively, in summer, when visitor numbers were viewed as too high to be tolerated because of impacts on elements such as parking and shopping (McFarland, 1978). This was a similar pattern to the behaviour of residents of Amsterdam (Gerritsma, 2019). Vargas-Sanches et al. (2014), in one of the few papers linking resident attitudes to seasonality, found that attitudes towards tourists and the impacts of tourism change between high and low seasons (Vargas-Sanches et al., 2014), with less resentment to tourists and tourism being present in the low season, as may have been expected.

In many cases the tourist summer season reflects not only when tourists are present in a destination but also when residents themselves are most likely to be out of their homes and in contact, visually and otherwise, with tourists and also to be in competition with them for space in car parks, shops, streets and recreation areas. It may be this disruption of 'normal' activities by tourists that drives some of the complaints about overtourism, rather than only the numbers of visitors present. In rural areas such disturbance and disruption may be more serious, even if absolute numbers are much lower than in urban destinations. Activities such as shopping can be done throughout the year and at different times of the week or day in most urban centres, but in rural areas the times and seasons are often of greater significance. Harvesting can be done only at specific times of the year, and often in even small windows such as specific days as crops ripen. Disruption to access because of traffic can be a major hurdle to such activities where road access is limited (Butler, 2019). Dealing with livestock can involve movement from one location to another at specific times of day (milking of cows for example), which may take place on foot using local roads, or on particular days to move stock to markets. There are attendant risks of competing with tourist traffic for road and ferry space, resulting in delays or missed connections. As tourist numbers are generally much lower in absolute terms in rural than urban destinations, seasonal fluctuations in tourist numbers may be of much higher significance in the former than the latter, and much more difficult to resolve where there are limited travel alternatives and little surplus capacity.

It is also likely that the relationship between overtourism and seasonality is linked to the level of development of the particular destinations. It is clear that overtourism has to be placed in the context of the degree of development of infrastructure in a destination and its capacity to absorb visitors. If visitor numbers regularly exceed the capacity of the infrastructure (such as water, accommodation, entertainment facilities, food and beverage operations, parking, and natural resources), then overtourism may well be the result and an appropriate description of the situation there, regardless of the 'season'. If numbers of visitors are below the capacity levels for most of the visitor season, then occasional overcrowding might be expected and generally tolerated, even if it is regular and repeated (for example on specific holidays as noted above) and the destination may be thought of as merely 'busy', even if numbers are very large.

In general, residents of established tourist destinations that have been specifically designed and created as resorts to cater to large numbers of visitors do not often complain about overtourism, as opposed to residents of primarily urban communities, such as Venice and Dubrovnik, that for reasons of history, location, and culture attract large numbers of tourists. Residents of heavily visited urban centres, however, which have or have had, other functions as their primary raison d'être, appear more likely to find tourists a nuisance, and less desirable than do residents in places whose primary function is tourism. Thus, overtourism reflects a number of variables such as original and current purposes of a destination, the capacity of the facilities for visitors, as well as the characteristics and behaviour of the visitors. Where the community involved is a combination of both a resort and a 'normal' urban centre, the threshold for acceptance of large numbers of visitors is likely to fluctuate greatly on a temporal basis, as large numbers of tourists will represent a change in the traditional pattern of urban life at specific times, ranging from a whole season to a few days during the season.

Case study: Venice – conflict and crowding

Venice is perhaps the most cited and extreme example of excessive tourist visitation and has been noted as such for many years, certainly as far back as the 19th century when John Ruskin complained of the presence of too many tourists. Venice warranted specific chapters in both Milano et al. (2019) and Dodds and Butler (2019), and many writers have blamed the problem on the lack of will to place any restrictions on visitor numbers, along with allowing cruise ships to dock at the port, worsening the problem in recent years. The number of day visitors exceed 35 million a year and overnight visitors another 10 million annually. There is little doubt that the pressure from tourism has been a factor in the reduction of the permanent resident population from almost 80,000 in 1990 to below 55,000 in 2018. The function and part of the fabric of the city have become dominated by tourism.

The pressure from tourists, however, while fluctuating seasonally, is of such a level overall that eleven of the twelve months of the year are deemed liable for a tourist tax, implying that overtourism in Venice is a regular and consistent issue. This viewpoint that overtourism in Venice is a persistent and almost permanent feature is supported by Visenti and Bertocchi (2019: 32) who note *"Venice, however, does not reflect a diversification of periods".* They go on to say that *"an off-season period would allow not only the recuperation of natural elements, but also a rest period for local communities and the maintenance and improvement of facilities"* (ibid). This would suggest that Venice is an exception to the general pattern, where overtourism is temporally sensitive. To some degree that situation is because many of the attractions of Venice, being the cultural heritage of the buildings and their artistic content, and the canals, are not as weather dependent as in some other destinations and thus worth visiting at any time of the year. It can be argued that the peak summer season is in many instances too hot and has unpleasant aromas from the canals, and the 'shoulder seasons' are more amenable to walking around the city.

Venice has in the past suffered from ineffective or non-existent control of tourism, exacerbated by the fact that, as in other places, (e.g. Barcelona), control of the means of access such as the airport, is not within its mandate. Thus, the numbers of tourists coming to visit Venice are outside its control. While few hotels have been established in Venice in recent years, hotel construction in Mestre and nearby districts has continued apace, allowing guests to visit Venice for the day, hence the very large number of day visitors noted above. Ironically, as an island, Venice should be able to limit and control visitors, both in terms of numbers and timing. Entry to the city is via the road and rail causeway, which would in principle be easy to control. Visitation could be limited to those holding permits (a Venice Card concept was proposed some years ago but never implemented).

In summer 2022 it was announced that new measures were to be introduced to control excess visitation (Citta di Venezia, n.d.). However, it is not clear at the time of writing (October 2022) how the controls will take place, as there are many exemptions for residents throughout the region and for many other types of visitors; nor how much they will cost to be implemented and maintained (Euronews, 2022). Marente (2022) has commented *"the ticketing system is just one of the initiatives to be realized (along with) new products and experiences for more responsible tourist targets; advanced mobility management; use of spaces (for residents and visitors); advanced and user-friendly information and booking tools; a different relationship with tourist intermediaries; and creativity and innovation of tourist operators. These inputs are certainly among the crucial elements to be developed – and should complement the introduction of the ticketing – so that there will be not only (just) control of tourism flows but also and above all, a positive evolution of the tourism and economic model of the city."*

Venice remains perhaps the ultimate problem, being vulnerable to easy cheap access from markets, a previous lack of control of the means of arrival of visitors and a lack of will to exert control over numbers which, combined with a world-renowned image and reputation has made it one of the top tourist attractions in the world. The result has been overtourism on an almost year-round non-seasonal basis. Whether the newly proposed measures, probably to be implemented throughout the year, will be effective in relieving overtourism remains to be seen. It has to be hoped that in this respect, Venice remains unique, that it is able to find the will and use the new measures to reduce the pressure on its fabric and its inhabitants.

Conclusions

It is hard to disagree that at least a part of the appearance of overtourism in most destinations coincides with the peak tourist season in those destinations. Overtourism and the annoyance and disruptions that it causes, are almost always recorded during the busiest time of the year at those destinations so affected. Solutions are not easy to envisage, as dispersing visitors to other times of the year may succeed only in increasing the length of time overtourism is experienced, and perhaps introduce it in times of year more sensitive to disruption than the peak summer season. Thus, while temporal influences are of significance in the occurrence of overtourism, temporal solutions may not be the answer. There have always been fluctuations in the level of visitation to destinations throughout the world, caused by both forces in the regions of origin of tourists and in the destinations. These include factors such as climate, institutional constraints, fashion and natural cycles (Butler 2001) and such variations on a temporal basis are likely to remain in the future, whatever the nature and level of tourism.

To avoid overtourism it is necessary to take a wider view than simply to extend the season and encourage 'out of season' visitation, as those measures would require much greater control over tourist flows, integration of other activities and development and extensive information provision and de-promotion of specific features at specific times. The tourist industry generally has shown little inclination to engage in such tasks, partly because of the element of competition (at national, regional, local and establishment levels), partly because of a refusal to accept that numbers are a major part of the problem of overtourism, and partly because of the difficulty of changing human preferences and desires. Unfortunately, it has taken something as catastrophic as the Covid-19 pandemic to temporarily halt overtourism, regardless of season. Permanent solutions are unlikely to be as effective and overtourism, like seasonality, is probably going to remain a feature of tourism in many destinations in the future.

Self-reflection questions

1. Is overtourism in Venice a product of lack of will of the authorities to limit numbers, or the unique nature of Venice as a tourist attraction?
2. What mitigation measures could be taken to ease the pressure on Venice?
3. Are the problems faced by Venice unique or are similar problems evident in other places?
4. Would it be acceptable to make Venice (or other places) too expensive to visit for most people in order to reduce visitor numbers?

References

Brougham, J.E. & Butler, R.W. (1981) The application of segregation analysis to explain resident attitudes to social impacts of tourism *Annals of Tourism Research*, **8**(IV), 569590.

Buhalis, D. (2020) There is no Overtourism - only badly managed tourism, *Travel Daily News*, https://www.traveldailynews.com/post/dimitrios-buhalis-bournemouth-university-there-is-no-overtourism-only-badly-managed-tourism (Accessed 12 Dec 2020)

Butcher, J. (2020) The War on Tourism, www.spiked-online.com/2020/05/04/the-war-on-tourism

Butler, R.W. (2001) Seasonality in tourism: Issues and implications, in Baum, T. & Lundtorp, S. (eds*) Seasonality in Tourism* Amsterdam: Pergamon, pp5-22

Butler, R.W. (2019) Overtourism in rural settings: the Scottish Highlands and Islands in Dodds, R. & Butler, R.W. (eds.) *Overtourism, Issues, Realities and Solutions* Berlin: De Gruyter, pp. 199-213.

Citta di Venezia, (n.d.) www.comune.venezia.it/it/content/contributo-accesso

Clawson, M. (1959) *The Crisis in Outdoor Recreation*, Baltimore: Johns Hopkins Press

Crux, R.G. & Legaspi, F.A.(2019) Boracy beach closure: the role of the government and the private sector, in Dodds, R. & Butler, R.W. (eds.) *Overtourism, Issues, Realities and Solutions* Berlin: De Gruyter, pp.95-110.

Darling, F.F. & Eichorn, N.D. (1967) *Man and Nature in the National Parks*, Washington,D.C.: Conservation Foundation

Dodds, R. & Butler, R.W. (2019) *Overtourism, Issues, Realities and Solutions* Berlin: De Gruyter

Dodds, R. & Butler, R. (2019), The phenomena of overtourism: a review, *International Journal of Tourism Cities*, **5** (4), 519-528.

Dredge, D. (2017) 'Overtourism' Old wine in new bottles? Linkedin, 13 Sept. www.linkedin.com/pulse/overtourism-old-wine-new-bottles-

Euronews, (2022) www.euronews.com/travel/2022/07/01/venice-sets-date-for-introduction-of-ticketing-and-entry-fees-heres-what-you-need-to-know

Gerritsma, R. (2019) Overcrowded Amsterdam: Striving for balance between trade, tolerance and tourism, in Milano, C., Cheer, J.M., & Novelli, M. (eds.) *Overtourism Excesses, Discontents and Measures in Travel and Tourism* Wallingford: CABI, pp. 125-148.

Horne, M. (2016) Cathedral bans tourists from funerals after selfies taken, p.22., *The Times* 10 August.

Lee C., Bergin-Seers, Galloway, G., O'Mahoney, B. & McMurray, A. (2008) *Seasonality in the Tourism Industry: Impacts and Strategies*, Gold Coast, Queensland: CRC for Sustainable Tourism

Marente, M. (2022) Personal communication, October, 2022.

McFarland, P. (1978) Attitudes of residents of Niagara Falls to visitor presence Unpublished Masters thesis, Department of Geography, University of Western Ontario, London.

Mihalic, T. (2020). Conceptualising overtourism: A sustainability approach. *Annals of Tourism Research*, 84, 103025.

Milano, C., Cheer, J., & Novelli, M. (Eds.). (2019). *Overtourism: Excesses, discontents and measures in travel & tourism.* Wallingford: CABI.

Pecot (M) & Ricaurte-Quijano, C. (2019) '¿Todos a Galápagos?' Overtourism in wilderness areas of the Global South. In C. Milan, J. M. Cheer & M. Novelli (eds.), *Overtourism: Excesses, discontents and measures in travel and tourism* (pp. 70-85). Wallingford UK: CAB International.

Peeters, P., Gössling, S., Klijs, J., Milano, C., Novelli, M., Dijkmans, C., Eijgelaar, E., Hartman, S., Heslinga, J., Isaac, R., Mitas, O., Moretti, S., Nawijn, J., Papp, B. & Postma, A., (2018), Research for TRAN Committee - Overtourism: impact and possible policy responses, European Parliament, Policy Department for Structural and Cohesion Policies, Brussels. https://www.europarl.europa.eu/RegData/etudes/STUD/2018/629184/IPOL_STU(2018)629184_EN.pdf

Tourism Geographies (2020) Special Issue, Volume 20 (3)

UN World Tourism Organization (UNWTO) Centre of Expertise Leisure, Tourism & Hospitality; NHTV Breda University of Applied Sciences;

and NHL Stenden University of Applied Sciences (2018), 'Overtourism'? – Understanding and Managing Urban Tourism Growth beyond Perceptions, Executive Summary, UNWTO, Madrid, DOI: https://doi.org/10.18111/9789284420070.

Vargas-Sanches, A., Porras-Bueno,N. & de los Angeles, M. (2014) Residents' attitude to tourism and seasonality, *Journal of Travel Research*. **53**(5), 581–596

Visentin, F. & Bertocchi, D. (2019) Venice: An analysis of tourism excess in an overtourism icon, in Milano, C., Cheer, J.M., & Novelli, M. (eds.) *Overtourism Excesses, Discontents and Measures in Travel and Tourism* Wallingford: CABI, pp18-38.

Wilkinson, P.F. (1996) Graphical images of the commonwealth Caribbean The tourist cycle of evolution, in Harrison, L.C. and Husbands, W. (eds.) *Practising Responsible Tourism: International Case Studies of Tourism Planning, Policy and Development* Toronto: John Wiley and Sons, pp.16-40.

18 Endnote: Covid and Beyond Covid: Temporal Futures

Philip Goulding

The chapters of this book were developed and written by the respective authors and contributors during the two years of the Coronavirus pandemic, in 2020-2021. Given the timing and timescale of the development of book's development, the purpose of this chapter is to provide a post-pandemic post-script for some of the key issues raised in the various chapters. It starts with a review of how the pandemic played out in terms of temporal patterns, followed by a discussion of recent developments that impinge on the relationship between temporality and tourism going forward.

The temporal hiatus of the Coronavirus pandemic

For much of the pandemic period, tourism was essentially 'on hold' in most parts of the world, with some tentative signs of activity during the periods of relaxation between total and partial lockdowns. Touristic activity resumed at different paces and at different timescales from country to country and region to region. According to UNWTO (2023), by 2022, Europe had reached around 80% of its pre-pandemic arrivals level, compared with just 23% in East Asia and 63% globally. The 900 million international visitors provisionally recorded that year represented a doubling of the previous year (UNWTO, 2023). The regional differences reflected the degree of relaxation of the state controls placed on human movement, on the relaxation of temporal trading restrictions for service-oriented businesses, on the recovery of transport infrastructures, capacities and schedules and on the degree of pent-up demand for each primary travel motivation (holidays, visiting friends and relatives, business contacts, pilgrimages).

The construct of temporality therefore seemed to take on a distinct meaning during the two years of the 'pandemic era'. For consumers, lockdown necessitated a re-appraisal of leisure time, where virtual or online 'armchair travel' was a reality for weeks or months on end (three months between March-June in the UK in 2020 and up to two years in parts of China and in Hong Kong), while more 'Covid-relaxed' approaches (Sweden and Australia during 2020) enabled 'real life' touristic activity to continue as before, albeit at reduced levels and with the realities of social distancing and service level modifications arising from stringent hygiene measures.

During lockdown, innovative tourism businesses adapted to 'real time' service encounters online. Museums and galleries across the world, including flagship attractions such as the Museum of Modern Art in New York and the British Museum took advantage of the opportunities of 'locked down consumers' to provide alternative experiences such as timed curated virtual tours and online exhibitions, children's entertainments, learning programmes, symposia and 'teach-ins' for potential customers via video conferencing and social media (ICOM, 2021; Leask, 2022). Innovative chefs and restauranteurs such as the Executive Chef of Wagamama (UK) did likewise, providing demonstrations on healthy eating or culinary skills (Independent, 2020). Timed events, online, brought a new temporal dimension to business. Apart from the social and wellbeing benefits of such innovations, they were perhaps cognisant of the importance of customer relationship management and reputational capital within the lockdown environment.

Meanwhile, the reduction of traffic on the ground and particularly in the skies led to a temporary drop in CO_2 emissions (NASA, 2021). IATA data show a reduction in global airline passengers from 4.8 billion in 2019 to 1.8 billion in 2020 (IATA, 2022). Light pollution reduced in many urban areas during this period (Gardiner, 2020) with night-time illuminations suspended, leading to a resurgence of interest in the night sky. Astrotourism was effectively happening in real-time, albeit from home or nearby, as opposed to within designated Dark Sky Reserves. Since the pandemic, the natural nocturnal phenomena of the Aurora Borealis and Australis have witnessed a surge in demand from consumers, which cruise operators in particular have been quick to exploit.

For the best part of two years, events and festivals ceased to be destination draws, as long as venues remained closed or operated with much reduced capacity. Sporting calendars were likewise suspended in much of the world during 2020 and part of 2021, meaning a temporary disappearance of short-term periodic peaks in visitation, hitherto characteristic of many urban areas' temporal tourism profiles. A major temporal shift in the mega-events cycle occurred with the re-timing of the summer Olympic Games in Tokyo postponed from 2020 to 2021 and executed in the absence of thousands of spectators who would otherwise have travelled to the event from around the world.

By the start of 2023, pandemic-related closures and social distancing requirements were mainly confined to China, Hong Kong and Macao. Chinese Golden Week in 2023 was nevertheless expected to witness 2.1 billion domestic journeys, around 30% lower than volume of travel compared to pre-pandemic years (Kyodo News, 2023), though still representing the world's largest temporal peak in travel generation.

Constructs and causation of temporality in tourism

In the era of post-pandemic recovery, there are early signs that the constructs of time and their inter-relationships with tourism are resuming to pre-pandemic patterns. It will, however, be a few years before post-Covid trends in traffic movements and visitor numbers, visitor spend, occupancy, load factors and other metrics enable analysts to detect signs of shifts in demand patterns. Nevertheless, as discussed in Chapters 1, 11 and 16, social trends are mitigating shifts in work-life balance in societies that generate much of the world's tourism demand. For example, the move towards shorter working weeks and greater flexibility in working time is gaining momentum and growing acceptance by employers (Figueira & Costa, 2022). A temporal implication of this is greater flexibility for short breaks through, for example, extended weekends. In Japan, Indonesia and elsewhere in Asia, there are signs of shifts in work patterns in the labour market to embrace 'workcations', defined as *"digital nomadism that seeks alternatives to existing work styles"* (Matsushita, 2022:1).

Demography also has a part to play in temporal shifts. The continuing demographic transition towards increasingly older populations in most post-industrial developed countries translates into a growth of potentially 'time rich' individuals. Where this coincides with financial security, disposable income and better health into older age, tourism businesses will be keen to exploit the possibilities of such segments, not least because of their propensity to travel 'off-season' or in periods of lower demand. Eurostat data indicates a more seasonally-spread pattern of visitation by 'older people' (55+) than the younger age groups, benefitting the shoulder months of May, September and October in particular (Eurostat, 2021).

A theme that is explicit in several of the chapters and implicit in others is the role of institutional factors in facilitating the temporal nature of tourism. As we saw in Chapter 2, such factors include statutory interventions in the form of public holidays and legislated holiday entitlements, and non-statutory but culturally fundamental 'institutions' in the form of religious celebrations or pilgrimages, festive seasons and so on. What happened during the pandemic period was a near-global set of interventions, predicated by public health protection, that extended the importance of institutional factors within the temporal mix. For example, when periodic Covid outbreaks occurred in Australia, snap lockdowns were instigated by the public authorities to bring case rates back to zero. The city of Perth shut down for five days in January 2021 when a single case was detected. Similarly, an outbreak of the Delta strain in Sydney in mid-June 2021 put Australia's largest city back into a lockdown that lasted several weeks (BBC, 2021). Enforced shutdowns limited human movements and face-to-face business operations. These in turn reinforced public anxieties among sectors of the populations over safety and certainly in touristic and leisure activities during periods of relaxation. 'Re-entry syndrome' has been coined as a term to capture the anxiety of re-engaging in activities in populated places (Van Niekirk, 2021). The interventions exemplify periodic disruption to the functioning of tourism.

On the supply-side, the challenges of recruitment to hospitality, aviation and visitor service organisations have intensified since the Covid pandemic, when furloughing and laying-off of staff became the norm. The challenges are particularly acute in cool temperate destinations such as the UK, Canada and parts of northern Europe with more pronounced employment seasons than in warm temperate destinations. The issues of recruitment and retention as outlined by the authors in Chapter 10 are ongoing realities in the UK hospitality sector. Combined with the current (since 2022) hike in wholesale energy prices and raw material costs across much of Europe, this labour squeeze has led to more commonplace reductions in trading days or trading hours for many hospitality, visitor attractions and destination service providers, signalling more fluidity in the short-term periodic supply of tourism services.

Climate, crowding and temporal remediation

Climate and the temporal spread of tourism are inextricably linked. The climate crisis is pervasive fuelling environmental degradation, natural disasters, weather extremes, food and water insecurity, economic disruption (UN, no date). Accordingly, any reassessment of tourism temporality should place climate at the centre of the debate.

As previously noted, there is evidence of a worldwide post-Covid tourism 'rebound', reflecting the release of pent-up demand for travel across all motivations. IATA data indicates, for example, a forecast growth in the region of 70% in air passengers in 2022 over the previous year and passenger numbers expected to surpass the four billion mark for the first time since 2019, despite ongoing geopolitical and economic uncertainties (IATA, 2022). Hence, tourism continues to exert pressures on the environment.

Nature-based tourism is particularly vulnerable to the impacts of climate induced weather events. The ongoing threat to habitats and species promotes 'last chance tourism', as examined in Chapter 3. Ironically, from a temporal perspective, this phenomenon exerts further pressures on those natural resources. On the Serengeti and Maasai plain in East Africa, the popularity of the 'Great Migrations' from June till September is similarly leading to overcrowding and environmental degradation, giving rise to concerns about limiting future market growth during those months (Conservation Action Trust, 2019).

The answer may lie in seasonal spread strategies, limiting visitor numbers and/or adopting seasonal pricing to regulate the demand for such experiences or, more controversially, closing the area for natural recovery. Thus, overtourism is no longer the preserve of Venice, Dubrovnik and other historic cities, as discussed in Chapter 17. One online travel advisory service exists to offer consumers help in finding the best time to visit popular attractions, using real-time crowd predictions and data-driven insights (avoid-crowds.com, nd.). 'Crowd-avoiding time' has therefore entered the framework of defining the visitor experience. Meanwhile, as the authors in Chapter 12 noted, 'seasonality mitigation' has become less overt as a policy issue for many destinations, rather subsumed within sustainability, even as the consequences of temporal imbalances remains live.

In conclusion, if 'time is of the essence', the essence of future paradigms for studying tourism will be to place greater emphasis on understanding its temporal dynamics.

Philip Goulding, February 2023

References

BBC (2021). Covid-19 rules: How six countries fared after easing Covid rules, BBC News, 15th July. https://www.bbc.co.uk/news/world-57796133

Conservation Action Trust (2019). Too much tourism in the Serengeti? https://conservationaction.co.za/resources/reports/too-much-tourism-in-the-serengeti/

Eurostat (2021). *Tourism Trends and Ageing*. Brussels: Eurostat. https://ec.europa.eu/eurostat/statistics-explained/index.php?title=Tourism_trends_and_ageing#Seasonal_patterns

Figueira, A. & Costa, S.R.R. (2022). Flexible arrangements as a trend on the future of work: a systematic literature review. *Revista S&G* 17, 2. https://revistasg.emnuvens.com.br/sg/article/view/1675

Independent (2020). Coronavirus: Wagamama launches free virtual 'Wok From Home' cooking lessons, 2nd April. https://www.independent.co.uk/life-style/food-and-drink/wagamama-coronavirus-virtual-cooking-lesson-chicken-katsu-a9441971.html

International Air Transport Association (2022). *Industry Statistics Factsheet*. https://www.iata.org/en/iata-repository/publications/economic-reports/airline-industry-economic-performance---june-2022---data-tables/

International Council of Museums (2021). *Museums, museum professionals and Covid-19: third survey*. https://icom.museum/wp-content/uploads/2021/07/Museums-and-Covid-19_third-ICOM-report.pdf

Kyodo News (2023) China expects 2.1 billion trips during New Year holiday season, 6th January. english.kyodonews.net/news/2023/01/53270fc9f55c-china-expects-21-billion-trips-during-new-year-holiday-season.html.

Leask, A. (2022). Visitor attractions and the Covid-19 pandemic, In Fyall. A., Garrod, B. Leask, A. & Wanhill, S. (eds.) *Managing Visitor Attractions*, 3rd edition. Abingdon, Oxon: Routledge. pp 445-449.

Matsushita, K. (2022). How the Japanese workcation embraces digital nomadic work style employees. *World Leisure Journal*.DOI: 10.1080/16078055.2022.2156594

NASA (2021). *Emission Reductions From Pandemic Had Unexpected Effects on Atmosphere*. NASA Jet Propulsion Laboratory, California Institute of Technology. 9th November. https://www.jpl.nasa.gov/news/emission-reductions-from-pandemic-had-unexpected-effects-on-atmosphere

Gardiner, B. (2020). Pollution made COVID-19 worse. Now, lockdowns are clearing the air. *National Geographic*, 8th April. https://www.nationalgeographic.com/science/article/pollution-made-the-pandemic-worse-but-lockdowns-clean-the-sky

United Nations (no date). The Climate Crisis – A Race We Can Win. UN75, http://www.un.org/UN75

UNWTO (2023). *Tourism Set To Return To Pre-Pandemic Levels In Some Regions In 2023*. UNWTO, January 17th. https://www.unwto.org/taxonomy/term/347

Van Niekirk, J. (2021). Take things slowly as lockdown ends to avoid 're-entry' syndrome. *The Guardian*, 13th March. www.theguardian.com/world/2021/mar/13/covid-take-things-slowly-lockdown-ends-avoid-re-entry-syndrome

Index

Accessibility 213
Advances in Destination Management (ADM) 159
Adventure tourism 44, 45
Affiliate marketing 197
Age of Discoveries (The) 57
AIDA Cruises 195
AirBnB 92, 136, 213
Airlines 86–88, 91, 99, 100, 101, 102, 209
Alpine Club 62
American Airlines 87, 91
Artificial Intelligence (AI) 91, 197, 213, 216
Astrotourism 45, 48, 235
Augmented Reality (AR) 213, 214
Aurora Australis / Southern Lights 77, 235
Aurora Borealis / Northern Lights 77, 235
Authoritarian regimes 63, 64
Average Daily rate (ADR) 96–98, 101

Babyboomer 39
Backpacking 52, 139
BajaBikes 196, 197
Bala Chaturdashi Festival 74
Berkshire Hathaway 115
Best Flexible Rate (BFR) 99
Big data 91, 197, 198, 210
Blockchain 214, 215
BOAC (British Overseas Airways Corporation) 86
Booking.com 100
British Museum 235
Budget airlines (*see also* low cost airlines) 87
Buffet, Warren 115
Burning Man Festival 170
Business Events 170
Byron, Lord George 58

Calendar effects 10
Capacity 89, 90, 91, 171 , 224, 228
Capacity planning 89, 192
Carter, Howard 60
Causal Factors 16-28
Chinese New Year 21, 22, 170, 225, 236
Christmas 20–22, 170, 183, 184, 186, 188
City Marketing Bureau 157
Climate 18, 21, 29-38, 108, 208, 238
Climate change 36–39, 208, 209
Climatic seasonality 17, 21, 29-42
Coachella Valley Music and Arts Festival 170
Commercial Home Operators 135, 136
Commonwealth Games 162
Competition and Markets Authority (CMA) 102, 103
Cook, Thomas 58, 59
Cook's Tours 59
CoolCousin 215
COP Climate Crisis Conferences 9
Coronavirus pandemic 19, 23, 113, 115, 156, 157, 195, 203, 204, 209, 220, 221, 230, 234-239
Crandell, Robert 86, 87
Crowd-avoiding time 238
Crown (The) television series 156
Cultural events 170

Dark sky reserves 48, 78
Dark sky tourism 48, 77, 78
Demand-derived influences 18-19
Demand forecasting 89, 90
Demographic trends 206, 207
Deregulation 86
Destination image 153-156, 162, 163, 175, 176, 185-188, 209

Destination Management Organisation (DMO) 103, 151-163, 174, 222
Destination Management Plan (DMP) 161
Destinations International 158
De-temporalisation 205-217
Dictatorships 63, 64
Digitization 113, 114, 197-200
DINAMO (Dynamic Inventory Allocation Modelling Optimizer) 86
Disintermediation 210–212
Disruptive innovations 212, 213
Diwali/Deepavali 22, 170
Dragon Boat Races (Hong Kong) 168, 169

Earl of Carnavon (The) 60
EatWith 213
Ecotourism 44, 48
Egyptology 60
Eid 188
Einstein, Albert 152
Elcano, Juan Sebastián 60
el destape 63
Employment, seasonal 119-130, 135, 139-144
Enlightenment (The) 57, 60
Enterprise-Resource-Planning (ERP) systems 199
European Capital of Culture 162
European City of Culture 172
Eurostat 104, 237
Event portfolios 173
Event tourism 166-179
Event typologies 168, 169
Expedia 100
Expeditions 60

Family businesses 134
Festivals 22, 74–76, 153, 167, 170–172, 175, 176
Festive season 21, 22
Fête de la Lumière 75
FIFA World Cup 9, 168
Fordism 64
Franco, Francisco 63
Friedrich, Kaspar David 62
Fruition factors 208, 209

Game of Thrones 154
Gaze, Henry 59
Gen X 39
Generation Y 134
Gen Z 39
German Alpine Association 193
Glastonbury Festival 170
Global Warming 36, 37
Golden Week (Japan/China) 20, 22, 225, 236
González de Clavijo, Ruy 60
Grand Tour (The) 57, 58, 66
Greater Miami Convention & Visitors' Bureau 35
Gregorian Calendar 22
Groupon 196

Hajj 21, 22
Halal tourism 207, 208
Handset culture 212
Hay Festival of Literature 170
Health tourism 57, 60, 61, 62
Hesse, Hermann 59
Hippie Trail (The) 65
Holi 22
Holographics 216
Hotels.com 100
Hurricane Katrina 154
Hygiénisme 61
IATA (International Air Transport Association) 209, 235, 238
IHG Group 99
Illumination typologies 74, 75
Industrial Revolution 20, 57, 60, 66
Institutional seasonality 19-22, 31, 32
Intermediation 100, 210–212
International Hotels Group (IHG) 99
International Year of Light 76

Julian Calendar 23
Jung, Carl Gustav 59

Kayak 211
Kumbh Mela 21, 22

Lantern Festival 22
Last Chance Tourism (LCT) 49, 50

Lastminute.com 100
Laterooms.com 100
Lévi-Strauss, Claude 59
Lifestyle 124, 125, 127, 139-145
Lifestyle business operators 24, 131-148
Lifestyle operator typologies 134, 135
Light 73-76
Light pollution 48, 78, 80, 235
'Littlewood's Rule' 86
Llum BCN 76
London Night Time Commission 78, 79
Lord of the Rings 154
Low cost airlines (*see also* budget airlines) 65, 87
Lunar New Year 20, 22
Lunn, Sir Henry 62

Magellan, Ferdinand 60
Marriott, J.W. Jnr 87
Marriott International 87
Mega-trends 206-209, 213-215
Miami Vice 35
MICE (Meetings, Incentives, Conferences, Exhibitions) 170
Migrant workers 120-128, 135-144
Millennials 39, 208
Mise en Lumière 77
Modifying actions 18, 21, 23
Mountain hikers 62
Museum of Modern Art (MOMA) 235

National Trust (The) (NT) 181, 184, 187
National Trust for Scotland (NTS) 7
Natural History Museum (London) 73
Natural migrations 47
Natural phenomena 47, 48
Nature tourism 43-55
New Orleans Convention & Visitors Bureau 154, 155
Nightlife licence 78
Nightscapes 48, 77, 78
Night-time recreation 72, 73
Niño (El) / Niña (La) 35, 36
Northern Lights / Aurora Borealis 77, 235
Notting Hill Carnival 153
Now Ruz 9

Occupancy 97, 98, 99, 101
Oktoberfest 170
Olympic Games 9, 153, 168, 169, 170, 172, 236
Online Travel Agent (OTA) 100, 102, 211, 212, 214
Overtourism 220-233, 238

Paris Agreement (The) 37
Pasteur, Louis 62
Paypal 200
PEOPLExpress 86, 87
Periodicity 9- 11, 13, 109
Perishability 87, 88, 96, 193
Planning process 107-118, 161, 162, 191-202
Pricing 85-94, 96-100, 102, 103, 192, 193
Personalised pricing 92
Price discrimination 89, 90
Polar bear viewing tourism 49
Post-Fordism 65, 66
Pride festivals 153
Public-private partnership (PPP) 158, 159
Push and pull factors 18, 29, 30, 31, 36

Recreational sports season 21
Re-intermediation 210–212
Revenue Generating Index (RGI) 97, 98
RevPAR (Revenue per available room) 96–99, 101
Revenue management (*see also* yield management) 86-91, 96, 98, 99
Richard III, King 155
Rosh Hashanah 10
Ruskin, John 62, 228
Ryanair 213

Schengen Agreement 207
Search engine advertising (SEA) 200
Search engine optimisation (SEO) 200
Seasonality 6-9, 12, 13, 119-128, 160-161, 206, 220-232
Social seasonality 19
Supply-side attributes 21-26
Seasonal langua ge (*see also* semiotics) 185-189
Seasonal workers 119-130

Seeker migrants 135
Semiotics 185-189
Seollal 22
Shoulder season 8
Ski-tourism 38, 39
Skyscanner.com 100, 211
Slowness / Slow Tourism 50-52
Smith Travel Research (STR) 96
Smithsonian magazine (The) 153
Social media 212
Son et Lumière 10, 75
Soviet Union 63, 64
Sports events 9, 153, 154, 160, 161, 169, 170, 236
STAR Report 96–100
Statutory holidays 20–22, 109
Steering logic 113-116
St. Lucy's Day 74
Stuttgart by Bike (SbB) 110-113, 193, 194, 196-201
Sub-arctic areas 34
Sukkot 188
Supply chain contracting 194-196
Supply chain management 192-196
Sustainability 223-226
Switzerland Tourism 157
Systems approach 5

Takapuna Beach Cup (New Zealand) 169
Temperate areas 33, 34
Temperature 32, 37, 38
Temporal marketing 180-190
Temporary Foreign Worker Program (Canada) 122, 125, 126
Terracotta Warriors 9
Tét (Vietman) 22
Thanksgiving 22
Thermalism 61
Thoreau, Henry David 62
Timurid Empire 60
Tourism Data Dashboard 104
Tourism in Difficult Areas (TiDA) 153, 158
Tour de France 154

Tour de Yorkshire 153
Tour operators 214
Tramlines Festival (Sheffield) 175, 176
Transnational workers 121-128, 135-144
Travelsupermarket.com 100
Trivago.com 100
Tropical areas 33
TUI 115, 199, 215

'Uncover Era' (*see* el destape) 63
United Airlines 91
Universal Declaration of Human Rights 64
UNWTO (United Nations World Tourism Organisation) 36, 104, 222, 234
Unusual weather patterns 35-38
Urban Tourism 72-74

Value chain management 113, 114, 191-202
VFR (visiting friends and relatives) 153
Virtual reality (VR) 197, 213, 214
VisitBritain 104, 155, 156
VisitEngland 104
Visit Isle of Man 161
Visit London 73

Wagamama 235
Weather 29-42, 109
Webjet (Australia) 215
Welcome to Yorkshire 12, 154
Whale Watching tourism 47
Wilderness tourism 44, 45
Wildlife tourism 44, 45, 46
Winding Tree 215
'workcations' 236
Working Holiday Visa (Canada) 122, 124, 125, 126
World Student Games 175
World Travel and Tourism Council (WTTC) 222

Yield management (*see also* revenue management) 86-91, 96, 98, 99
Yom Kippur 10

Place name Index

Afghanistan 60
Alaska 32
Amazon (The) 46
Amsterdam 222, 224, 225, 226
Antarctica 47
Athens 155
Auckland 169
Australia iii, 31, 32, 36, 47, 77, 124, 215, 235, 237
Austria 8, 185, 186

Balearic Islands (Spain) 23
Bali (Indonesia) 32
Barcelona iii, 58, 72, 75, 76, 153, 154, 221, 229
Beijing 11, 33
Benidorm (Spain) 64
Bhutan 23
Biarritz (France) 61
Blackpool (England) 74, 153
Black Forest (Germany) 115
Boracay (Philippines) 223
Bornholm (Denmark) 23
Brazil 8, 207
British Columbia (Canada) iv, 84, 120, 123, 124
British Museum (London) 235
Brittany (France) 188
Brooklyn Bridge (New York) 74
Bulgaria 8

Calke Abbey (England) iv, 181–184, 189
Cambodia 32, 33
Canada iii, iv, 32, 49, 120, 122–124, 126, 133, 237
Cap de Barbaria (Ibiza) 77
Cape Town 31
Caribbean 6, 21
Carinthia (Austria) 186
Carolinas (USA) 23
Chamonix (France) 31, 62, 132, 139–144
China 33, 207, 225, 235, 236
Churchill (Canada) 49
Clayoquot Sound, BC (Canada) 123

Cook Islands 32
Copenhagen 162
Cornwall (England) 133, 142
Crete 23
Croatia 8, 23, 154
Cuba 49
Czechia / Czech Republic 8

Dalarna (Sweden) 127
Delaware (USA) 156
Derbyshire (England) 181
Dominican Republic 8
Dubai 31
Dubrovnik 221, 227, 238

Edinburgh 175, 221
Egypt 58, 60
Eiffel Tower (Paris) 74, 185
Empire State Building (New York) 77
England 12, 48, 51, 104, 105, 160, 175, 181
European Alps 6, 57, 62

Fiji 22
Finland 8
Florida 6, 35
France 18, 31, 34, 58, 61, 132, 139, 154, 185

Galapagos Islands 46, 222
Galloway Forest Park (Scotland) 48
Glasgow (Scotland) 162
Goa (India) 31
Grand Canyon (Arizona, USA) 77
Great Barrier Reef (Australia) 45
Greece 58, 77

Highlands and Islands (Scotland) 24
Himalayas 36
Huaraz (Peru) 31
Hong Kong 102, 156, 168, 235, 236
Hudson Bay (Canada) 49

Ibiza / Eivissa 77
Iceland 32, 154
India 22, 65, 207

Indian Ocean 21
Indonesia 236
Iraq 60
Ireland 133
Isle of Man 160, 161
Italy 58

Japan 8, 20, 33, 47, 236
Jordan 155, 213

Kauai (Hawaii) 32
Kenya 238
Kielder Water (England) 48
Komodo Island (Indonesia) 208
Kumbu (Nepal) 31

Lapland (Finland) 34
Leicester (England) 58, 59, 154, 155
Leicestershire (England) 140, 141
Liverpool (England) 155
London iii, 8, 33, 59, 62, 72, 73, 75, 78, 79, 102, 142, 143, 153, 156
London Heathrow Airport 102
Loughborough (England) 58
Lyon (France) 75

Maasai Mara (Kenya) 238
Macao 236
Malaysia 32
Mallorca 115
Manchester (England) 102
Manitoba (Canada) 49
Mauritius 22, 31
Maya Bay (Thailand) 208
Mecca 22
Mediterranean 6, 31, 32, 34, 36, 37, 39, 40, 64, 65, 188
Melbourne (Australia) 75
Mestre (Italy) 229
Mexico 8, 24, 126
Miami iii, 31, 35
Miami Beach 35
Montsec Observation Centre (Spain) 77
Morzine (France) 34
Moscow 33
Museum of Modern Art (MOMA) (New York) 235

Nashville (USA) 155
Natural History Museum (London) 73
Nepal 31, 36, 65, 74
Netherlands 8
New England (USA) 47
New Orleans (USA) 154, 156
New York 8, 33, 74, 77, 78, 235
New Zealand 31, 32, 133, 154, 169
Niagara Falls 226
Ningaloo Coast (Australia) 47
North Korea 63
Northern Ireland 154, 181
Northumberland 48, 133
Northumberland National Park (England) 48
Norway 34

Ontario (Canada) 222
Orkney Isles (Scotland) 225

Pacific Ocean 35, 36, 123
Pacific Rim 123, 216
Pacific Rim National Park 123
Palestine 58
Paris 33, 37, 73, 74, 185, 221
Perth (Australia) 237
Peru 31
Petra (Jordan) 155
Philippines 36, 126, 223
Phu Quoc Island (Cambodia) 33
Portugal 58
Prague 221
Pyrenees 6, 77

Queensland (Australia) 32, 77
Queenstown (New Zealand) 31

Rio de Janeiro 31
Rocky Mountains 6

Saint Lucia 156
Sälen (Sweden) iv, 120, 127, 128
San Sebastian (Spain) 61
Sant Feliu de Guíxols (Spain) 61
Santorini (Greece) 77
Scandinavia 32, 74
Scotland 7, 20, 24, 32, 48, 52, 143

Serengeti (Tanzania) 238
Seychelles 32
Sheffield (England) iv, 149, 175, 176
Shetland Isles (Scotland) 32
Siberia (Russia) 32
Sicily 31
Singapore 33
South Korea 20, 22
Spain 61, 63, 64, 77
Stuttgart (Germany) iv, 5, 110, 111, 193, 194, 201
Sweden iv, 84, 120, 127, 235
Switzerland 18, 31, 157, 188
Sydney 237

Thailand 7, 8, 32, 208
Ticknell, Derbyshire (England) 181
Tignes (France) 31
Tofino (Canada) iv, 120, 123–126, 128
Tokyo 236
Trocadero Square (Paris) 74

Uganda 31
UK / United Kingdom iii, 4, 8, 12, 19, 22, 23, 48, 102, 104, 124, 136, 139, 142, 158, 173, 181, 215, 235, 237
USA / United States of America 21–23, 31, 35, 36, 47, 58, 64, 77, 86, 188, 207

Vancouver Island (Canada) 120, 123
Venice iv, vi, 5, 204, 221, 224, 225, 227–231, 238
Vermont (USA) 188
Victoria Harbour (Hong Kong) 169
Vienna 186, 188
Vietnam 22, 32

Wales 181
Western Australia 47
Yorkshire iii, 12, 104, 153, 154
Zermatt (Switzerland) 31

www.ingramcontent.com/pod-product-compliance
Ingram Content Group UK Ltd.
Pitfield, Milton Keynes, MK11 3LW, UK
UKHW050457150426
5217IPUK00025B/1728